Chemical Sensors

I0051153

Dnyandeo Pawar
Departmental Faculty of Engineering
Università Campus Bio-Medico di Roma
Rome, Italy

CRC Press
Taylor & Francis Group
Boca Raton   London   New York

CRC Press is an imprint of the
Taylor & Francis Group, an **informa** business
A SCIENCE PUBLISHERS BOOK

Cover image provided by the author.

First edition published 2024
by CRC Press
2385 NW Executive Center Drive, Suite 320, Boca Raton FL 33431

and by CRC Press
4 Park Square, Milton Park, Abingdon, Oxon, OX14 4RN

© 2024 Taylor & Francis Group, LLC

CRC Press is an imprint of Taylor & Francis Group, LLC

Library of Congress Cataloging-in-Publication Data (applied for)

ISBN: 978-0-367-86174-2 (hbk)
ISBN: 978-1-032-57353-3 (pbk)
ISBN: 978-1-003-02015-8 (ebk)

DOI: 10.1201/9781003020158

Typeset in Palatino Linotype
by Shubham Creation

Preface

Chemical sensor is a small-size device consisting of a transduction element and recognition element, capable of providing real-time sensing information related to chemical interaction and the surrounding environment of an analyte in complex samples. At present, the chemical sensing field has been gaining a rapid pace due to its massive demands in numerous applications, such as environmental monitoring, marine environments, healthcare sector, agriculture field, and other related industries. Recent developments in the field of nanotechnology have changed the complete scenario of chemical sensing and therefore, it is possible to develop a smart chemical sensor. Numerous nanostructures of several materials including semiconductor metal oxides, 2D, and polymers have been extensively used in chemical sensing. Simple and low-cost synthesis methods have attracted considerable attention in the chemical sensing domain as, they not only produced versatile materials but also supported in mass-production of sensors. Present technology targets the design of new kinds of advanced and sophisticated thin-films deposition methods and therefore, it is possible to fabricate nano, micro-scale, and multi-functional sensory platforms for chemical sensing. Today, the world is looking at a smart chemical sensing platform, which could assist in monitoring human health and our ecosystem in a better way.

Chemical sensor aims to explore the comprehensive and integrated view of numerous chemical sensors based on new versatile materials, advanced thin-films deposition methods, newly designed chemical sensing approaches, and their applications towards the detection of various chemical moieties and future perspectives. The book includes industrial applications and will minimize the gap between the laboratory and real-time applications. It contains information about multidisciplinary fields, like material science, chemistry, physics, and engineering and therefore, it will be of interest to undergraduate and postgraduate students, faculty, professionals, engineers, and academic researchers. The book consists of the following nine chapters:

Chapter 1 summarizes the chemical sensing parameters. It emphasizes the structure of a chemical sensor and common parameters to analyze a chemical sensor. It also a discussion on the influence of major parameters on chemical sensing behavior.

Chapter 2 focuses on the mechanism of various chemical detection methods, which are extensively adopted in chemical sensing. Certain widely used techniques, like optical fiber, chemiresistive, electrochemical, capacitive, field effect transistors, and quartz crystal microbalance are thoroughly discussed. The control parameters of each detection method along with the possible scope to improve the performance are also discussed.

Chapter 3 presents mostly utilized materials, such as semiconducting metal oxides, 2D materials (graphene and its derivatives, transition metal chalcogenides, Mxenes, and metal organic frameworks), and polymers for chemical sensing. Their details regarding the importance in chemical sensing, synthesis methods along with their limitations, and futuristic material designs are discussed.

Chapter 4 discusses some broadly used advanced deposition thin-films techniques for chemical sensor development. The general working principle along with their role and fabrication parameters are discussed. Also, the advantages and disadvantages of each deposition technique are highlighted. This chapter provides a precise discussion of the physics and chemical principles behind these deposition techniques.

Chapter 5 reviews the most recent gas sensor development based on various extensively used methods, such as electrochemical, capacitive, chemiresistive, field effect transistor, fiber optics, and quartz crystal microbalance. Also, the possible scope in the advancement of new kinds of gas sensing approaches along with the smart materials structures and strategies are briefly summarized. The limitations of each method along with the possible scope for improvement are also presented.

Chapter 6 deals with various new and advanced methods, like electrochemical, field effect transistors, optical fiber, fluorescence, and colorimetric for ion sensing. This chapter gives the current state of the art and performances of these ion detection techniques. Future scope of the smart ion sensor development based on advanced materials fabrication is also highlighted.

Chapter 7 provides an understanding of the different techniques, such as electrochemical, capacitive, resistive, fiber optics, luminescent, colorimetry, and quartz crystal microbalance combined with extraordinary material structures for humidity sensing. The future tasks linked to the progress of humidity sensors are also discussed.

Chapter 8 focuses on the recent development in the field of pH sensing implemented by using numerous advanced techniques. This chapter also presents a new types of materials and their conduction mechanism and microstructure characteristics of the pH sensors. It also highlights the perspectives for future needs in fundamental and applied aspects of pH sensing.

Chapter 9 reviews numerous mostly used and highly compatible techniques, such as electrochemical, field effect transistors, fiber optics, whispering gallery mode, colorimetry, and fluorescence for biosensing. The discussion regarding materials, analyte, detection range, and limit of detection is discussed. The future outlook of the biosensing field is also outlined.

Dnyandeo Pawar

Contents

Chapter 1

Chemical Sensor Parameters

1.1 INTRODUCTION

The most powerful sensory system is present in every organism living on the earth. They interact with their surroundings through a powerful mechanism. These natural systems are based on 3'S' parameters, i.e., sensitive, selective, and stable. These systems are continuously inspiring us to design and develop a new systems with advanced features which can measure the physical and chemical parameters even in a complex environments.

The identification, quantification, and examining of chemical compounds in dissimilar surroundings is really a challenging task. In this case, the development of a simple, portable, ease in fabrication, low-cost, and sensitive sensory system is always required. As defined by the IUPAC, a chemical sensor is expressed as "a device that transforms chemical information, ranging from the concentration of a specific sample component to overall composition analysis, into an analytically useful signal" (Hulanicki et al. 1991). Chemical sensors are also defined as miniaturized analytical devices that can deliver real-time and online information in the presence of specific compounds or ions in complex samples (Wang and Wolfbeis 2020). In chemical sensing, online monitoring is a very important aspect. Also, the term 'sensor' needs to be used properly in chemical sensing domain. For example, the conventional cuvette tests exclusive of any (online) monitoring ability are considered as sensors, and thus, need to be addressed. Similarly, in the case of a biosensor, if there is no biological constituent (like enzymes, immunosystems, DNA/RNA, tissue or whole cells) present, it won't be considered a biosensor. For example, measurement of refractive index of glucose solution or pH of blood is not considered as biosensing (Wang and Wolfbeis 2016; 2020). Overall, a clear understanding of the 'sensor' term and its correct implementation in chemical sensing domain is extremely important.

Chemical sensor (Figure 1.1) gives the information regarding chemical nature and its surrounding environment. Generally, it consists of two parts, a transducer and a recognition element. The role of recognition element is to interact with an analyte or parameter of interest and produce a measurable signal via transduction. When it interacts with the analyte, it leads to changes in the material properties like mass, optical, and electrical. After that the role of a transducer comes into play, which transforms this chemical information into a readable form, mostly into an electrical signal. The sensing layer is very important and it decides the sensor response, selectivity, temporal characteristics, repeatability, and stability (Meng et al. 2019; Paolesse et al. 2017; Wang and Wolfbeis 2020). This chapter highlights the important parameters of a chemical sensor, and their role in deciding the chemical sensor's performance.

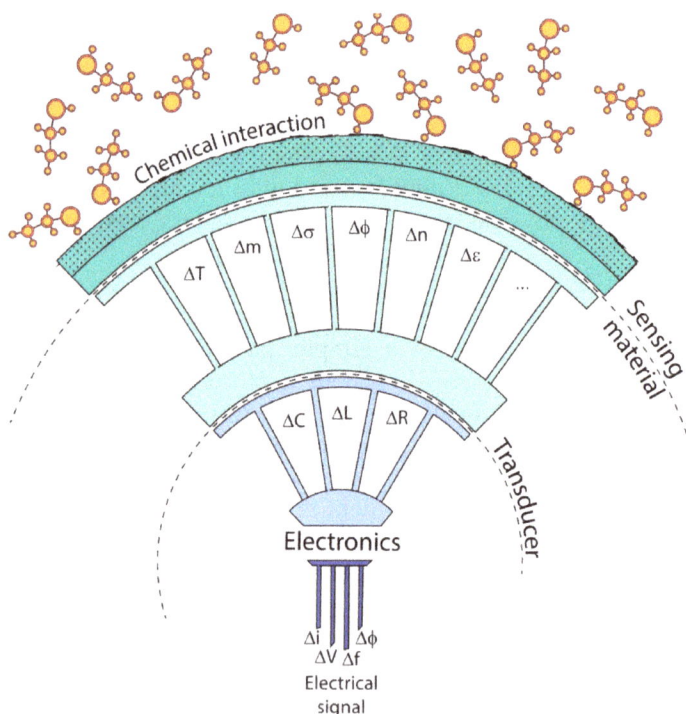

Figure 1.1 Common structure of a chemical sensor. The interaction of an analyte with the recognition element produces information that is further transduced by the transduction element into a readable form. [Reprinted with permission from Ref. (Paolesse et al. 2017); Copyright 2017, American Chemical Society].

1.2 COMMON PARAMETERS TO ANALYZE A CHEMICAL SENSOR

Chemical sensor can detect a parameter of interest qualitatively and quantitatively present in the environment. The chemical information may be collected from the chemical reaction or physical properties of the analyte. The following are the key parameters (shown in Figure 1.2) for a chemical sensor such as response, sensitivity, selectivity, limit of detection (LOD), response and recovery time, and long-term stability.

1. **Sensor response:** This is very crucial parameter of a chemical sensor used for real time monitoring applications. The chemical sensor response is given in terms of various parameters depending upon the type of configuration. For example, in the case of a chemiresistive-based sensor, the response is measured in terms of change in resistance. In amperometry chemical sensors, the sensor response is mentioned in terms of change in current. For inductive-based chemical sensors, the response is given in terms of change in inductance, etc. Generally, it is defined as the specific change in the physical parameter in presence of a change in concentration of analyte (Yoon et al. 2021).

2. **Sensor sensitivity:** It is the difference in measured signal per unit analyte concentration. Generally, it is the slope value (Figure 1.2a) and can be easily extracted from the calibration curve (Paolesse et al. 2017; Yoon et al. 2021).

3. **Selectivity:** This is a very vital parameter for a chemical sensor. It is the sensor's capability to detect a specific analyte among the various entities with a good response. To obtain this parameter, the sensor's response is measured with various analytes which might interfere with the signal. Thus, in this way the sensor's response is measured to each analyte and the best response is analyzed for a specific analyte only (Barik and Pradhan 2022).

4. **Detection limit:** It is the least concentration of an analyte that a sensor can detect above the noise signal as depicted in Figure 1.2c (Cao et al. 2022; Committee 1987).

$$\text{Detection limit} = 3 \times \frac{\text{Standard deviation of the response}}{\text{Sensitivity}} \quad (1.1)$$

Generally, the detection limit should be more than 3 times to that of noise signal present in the measured signal without analyte concentration.

5. **Limit of quantification:** It is the smallest analyte concentration that can be quantitatively sensed with a specified accuracy and

precision. Generally, it is ten times to noise level of the response (Cao et al. 2022; Krupčík et al. 2015).

$$\text{Limit of quantification} = 10 \times \frac{\text{Standard deviation of the response}}{\text{Sensitivity}} \quad (1.2)$$

6. **Response and recovery time:** It is a very important characteristics of a chemical sensor. The response time is the time taken by the sensor to get from 0% response value to 90% of its total response value. The recovery time, on the other hand, is the time taken by the sensor to go from 100% of its total response to 10% of its total response (Figure 1.2b) (Cao et al. 2020; Yao et al. 2021). For real time monitoring applications, the response/recovery time of a chemical sensor should be very fast. However, this may not be useful for a one-time used chemical sensor.

7. **Stability:** It is the measure of sensor performance over the period (Figure 1.2d). Generally, it is the capability of a sensor to generate reproducible results over a period-of-time (Paolesse et al. 2017; Yao et al. 2021). The time-period may be in minutes, days or even longer depending upon the materials' properties and impact of extrinsic parameters such as humidity and temperature. However, for a chemical sensor, the sensor should be long lasting without hampering its sensing performance.

8. **Repeatability:** Repeatability implies how the sensor shows the measured value more accurately (Paolesse et al. 2017; Yoon et al. 2021). This is the random error of series of measurements (Figure 1.2d). Generally, in the measurements, adding an error bar calculated from the series of experiments is always scientifically advisable.

9. **Reversibility:** It is the ability of a sensor to return to its baseline value on removal of analyte (Das et al. 2022; Yoon et al. 2021).

Ideally a chemical sensor should possess high sensitivity, selectivity, dynamic range and stability, low detection limit, good linearity, small response and recovery time, long-life cycle and should be able to produce the same output under same environmental conditions.

1.3   INFLUENCE OF MAJOR PARAMETERS ON THE SENSING BEHAVIOR

Nanotechnology has changed the complete scenario of chemical sensing. With advancements of new deposition techniques and methods,

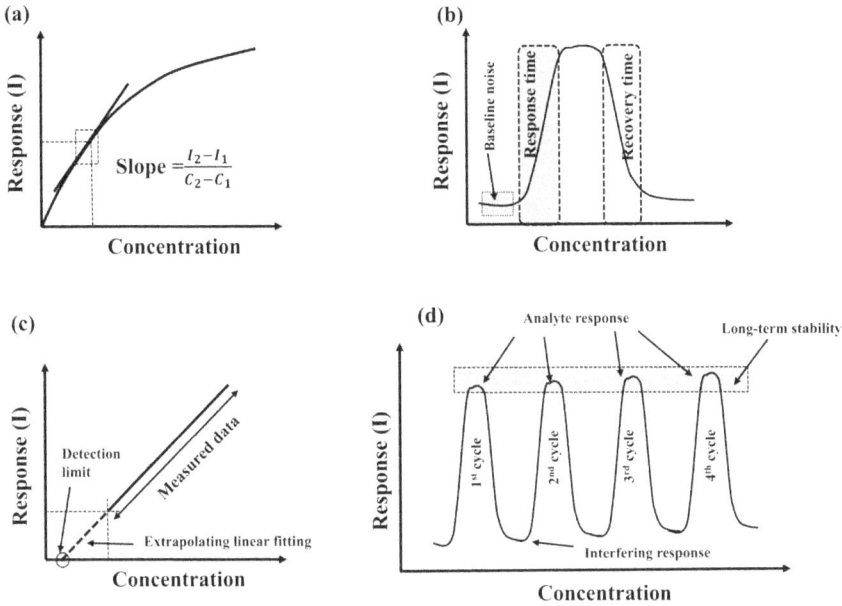

Figure 1.2   Different parameters of a chemical sensor.

miniaturized chemical sensors can be realized. Numerous synthesis methods have been utilized to prepare various materials such as metal oxides, carbon materials, metals, and polymers. These materials have proved to be very effective in chemical sensing domain due to their excellent physical and chemical properties. The details of these types of materials are provided in the subsequent chapter.

In chemical sensing, the sensing material plays a vital role. The output signal is generated through a well-known sensing mechanism. However, depending upon the materials, there are several other parameters (Figure 1.3) that could influence the sensing mechanism and decide the chemical sensor performance. The following discussion is provided for semiconductor metal oxides materials which are widely adopted for chemical sensing.

1. **Particle size, crystallite and grain size:** Generally, in the case of metal oxides sensing layer, these parameters affect the sensing performance. It is observed that the sensitivity of metal oxides can be improved by reducing the particle size. After decreasing the particle size, the surface energy gets increases and hence, improves the sensitivity (Das et al. 2022; Majumdar 2015; Rao et al. 2002; Xu et al. 1991). It also significantly improves the temporal characteristics of the chemical sensor. Similarly, the

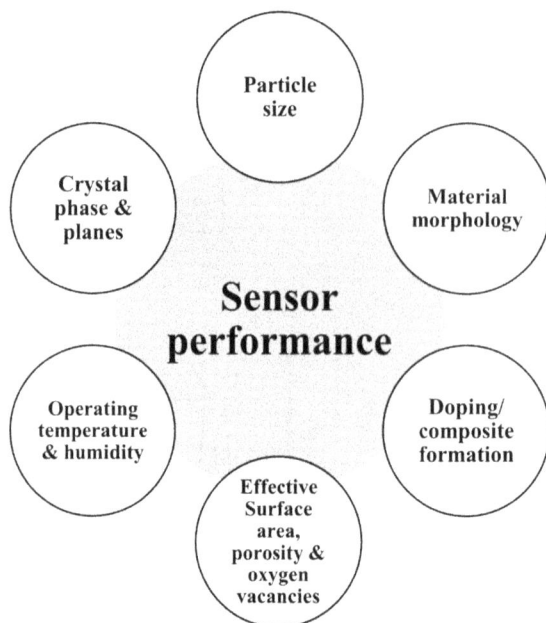

Figure 1.3   Numerous influential factors deciding the chemical sensor
performance.

grain size also affects the sensitivity as studied elsewhere.
Generally, the sensitivity is inversely proportionate to grain
size (Lu et al. 2000; Majumdar 2015).

   Also, the neck between two nearby particles plays a crucial
role in deciding the sensitivity. In general, when the particle
size increases, the effective contact zone among particles per
unit volume decreases and hence, reduction in sensitivity
(Korotcenkov 2007; Wan et al. 2014).

2. **Morphology:** The sensing performance of the sensor gets
   improved with respect to rise in quantum confinement. For
   example, from 3D to 1D structure (Gao and Zhang 2018; Li et al.
   2021; Pilliadugula and Gopalakrishnan 2022). The hollow spheres
   have higher a possibility to improve the sensing performance.
   The surface area will also change with reference to the material
   morphology and hence, it is a major influencing factor in
   chemical sensing.

3. **Specific surface area and porosity:** Highly porous materials
   exhibit great specific surface area and offer more active positions
   for adsorption of analyte and boost the sensing performance.
   In the case of gas sensing, different gases have different

diffusivities and hence, can easily reach through the pores and could enhance the sensor response (Galstyan et al. 2022; Hassan et al. 2021; Zairi et al. 2001). Generally, the specific surface area is inversely proportional to the crystallite size and density of the crystal. Thus, as the crystallite size decreases, there is an increase in specific surface area and hence, an increase in the chemisorption on the sensor surface (Das et al. 2022).

4. **Oxygen vacancy:** This is the most vital parameter and mainly depends upon the series of physical/chemical activities such as alteration in Fermi level, bond break and distorted lattice. The presence of oxygen vacancy accelerates the carrier transport rate and hence, enhances the sensor response (Al-Hashem et al. 2019; Zhang et al. 2020).

5. **Operating temperature and humidity:** The response of a chemical sensor can be effectively modulated by selecting the proper working temperature of the material. Generally, the sensor response follows 'increase-maximum-decay' nature. As the temperature increases, the surface activity also rises and attains the maximum curve but decreases further due to low adsorption energy with an increase in temperature. At a low temperature, small number of adsorbed oxygen species are available. At a higher temperature, there is a great chance of oxygen vacancies getting oxidized and thus, reduction in the concentration of oxygen vacancies present (Cao et al. 2022; Cao, et al. 2021a; Singh et al. 2019). Humidity also impacts the sensor performance by decreasing the base resistance. The water gets adsorbed at the vacant sites of the material, and therefore, decreases the adsorption sites for the analyte (Cao, et al. 2021b; Korotcenkov et al. 2007; Tischner et al. 2008).

There are many other parameters such as doping, high-index crystal facets, formation of heterojunction, homojunction, and oxides, etc., which could also impact the chemical sensor performance (Das et al. 2022; Franke et al. 2006; Korotcenkov 2007; Okeke et al. 2022; Prakash et al. 2021; Yamazoe et al. 1983). So, for the development of a smart chemical sensor, most of these parameters need to be considered.

1.4 CONCLUSION

In summary, a detailed analysis of all chemical sensing parameters is considered for real-time detection. The most influencing factors which might impact the sensing performance are also highlighted. The chemical sensor performance can be improved by adopting several strategies like adjusting the film thickness, tuning surface morphology, varying surface area by implementing different techniques, e.g., doping, composite formation, etc. Effect of certain environmental factors such as humidity and temperature on sensing performance also need to be investigated. Numerous deposition techniques are also important to decide the film's uniformity. Other influential factors and key points are addressed in subsequent chapters devoted to each chemical parameter detection.

REFERENCES

Al-Hashem, Mohamad, Sheikh Akbar and Patricia Morris. 2019. Role of oxygen vacancies in nanostructured metal-oxide gas sensors: a review. Sensors and Actuators B: Chemical 301: 126845. doi:https://doi.org/10.1016/j.snb.2019.126845.

Barik, Puspendu and Manik Pradhan. 2022. Selectivity in trace gas sensing: recent developments, challenges, and future perspectives. Analyst 147(6): 1024–1054. The Royal Society of Chemistry. doi:10.1039/D1AN02070F.

Cao, PeiJiang, YongZhi Cai, Dnyandeo Pawar, S.T. Navale, Ch.N. Rao, Shun Han, WangYin Xu, et al. 2020. Down to ppb level NO_2 detection by ZnO/rGO heterojunction based chemiresistive sensors. Chemical Engineering Journal 401: 125491. doi:https://doi.org/10.1016/j.cej.2020.125491.

Cao, PeiJiang, XingGao Gui, Dnyandeo Pawar, Shun Han, WangYing Xu, Ming Fang, XinKe Liu, et al. 2021a. Highly ordered mesoporous V_2O_5 nanospheres utilized chemiresistive sensors for selective detection of xylene. Materials Science and Engineering: B 265: 115031. doi:https://doi.org/10.1016/j.mseb.2020.115031.

Cao, PeiJiang, RongGuan Chen, YongZhi Cai, Dnyandeo Pawar, Ch.N. Rao, Shun Han, WangYin Xu, et al. 2021b. Ultra-high sensitive and ultra-low NO_2 detection at low-temperature based on ultrathin In_2O_3 nanosheets. Journal of Materials Science: Materials in Electronics 32(14): 19487–19498. doi:10.1007/s10854-021-06467-4.

Cao, PeiJiang, YongZhi Cai, Dnyandeo Pawar, Shun Han, WangYing Xu, Ming Fang, XinKe Liu, et al. 2022. Au@ZnO/rGO nanocomposite-based ultra-low detection limit highly sensitive and selective NO_2 gas sensor. Journal of Materials Chemistry C 10(11): 4295–4305. The Royal Society of Chemistry. doi:10.1039/D1TC05835E.

Committee, Analytical Methods. 1987. Recommendations for the definition, estimation and use of the detection limit. Analyst 112(2): 199–204. The Royal Society of Chemistry. doi:10.1039/AN9871200199.

Das, Sagnik, Subhajit Mojumder, Debdulal Saha and Mrinal Pal. 2022. Influence of major parameters on the sensing mechanism of semiconductor metal oxide based chemiresistive gas sensors: a review focused on personalized healthcare. Sensors and Actuators B: Chemical 352: 131066. doi:https://doi.org/10.1016/j.snb.2021.131066.

Franke, Marion E., Tobias J. Koplin and Ulrich Simon. 2006. Metal and metal oxide nanoparticles in chemiresistors: does the nanoscale matter? Small 2(1): 36–50. John Wiley & Sons. doi:https://doi.org/10.1002/smll.200500261.

Galstyan, Vardan, Abderrahim Moumen, Gayan W.C. Kumarage and Elisabetta Comini. 2022. Progress towards chemical gas sensors: nanowires and 2D semiconductors. Sensors and Actuators B: Chemical 357: 131466. doi:https://doi.org/10.1016/j.snb.2022.131466.

Gao, Xing and Tong Zhang. 2018. An overview: facet-dependent metal oxide semiconductor gas sensors. Sensors and Actuators B: Chemical 277: 604–633. doi:https://doi.org/10.1016/j.snb.2018.08.129.

Hassan, Israr U., Hiba Salim, Gowhar A. Naikoo, Tasbiha Awan, Riyaz A. Dar, Fareeha Arshad, Mohammed A. Tabidi, et al. 2021. A review on recent advances in hierarchically porous metal and metal oxide nanostructures as electrode materials for supercapacitors and non-enzymatic glucose sensors. Journal of Saudi Chemical Society 25(5): 101228. doi:https://doi.org/10.1016/j.jscs.2021.101228.

Hulanicki, A., S. Glab and F. Ingman 1991. Chemical sensors: definitions and classification. Pure and Applied Chemistry 63(9): 1247–1250. doi:doi:10.1351/pac199163091247.

Korotcenkov, G. 2007. Metal oxides for solid-state gas sensors: what determines our choice? Materials Science and Engineering: B139(1): 1–23. doi:https://doi.org/10.1016/j.mseb.2007.01.044.

Korotcenkov, G., I. Blinov, V. Brinzari, and J.R. Stetter. 2007. Effect of air humidity on gas response of SnO_2 thin film ozone sensors. Sensors and Actuators B: Chemical 122(2): 519–26. doi:https://doi.org/10.1016/j.snb.2006.06.025.

Krupčík, Ján, Pavel Májek, Roman Gorovenko, Jaroslav Blaško, Robert Kubinec and Pat Sandra. 2015. Considerations on the determination of the limit of detection and the limit of quantification in one-dimensional and comprehensive two-dimensional gas chromatography. Journal of Chromatography A 1396: 117–30. doi:https://doi.org/10.1016/j.chroma.2015.03.084.

Li, Zhong, Zhengjun Yao, Azhar Ali Haidry, Yange Luan, Yongli Chen, Bao Yue Zhang, Kai Xu, et al. 2021. Recent advances of atomically thin 2D heterostructures in sensing applications. Nano Today 40: 101287. doi:https://doi.org/10.1016/j.nantod.2021.101287.

Lu, Fan, Ying Liu, Mei Dong and Xiaoping Wang. 2000. Nanosized tin oxide as the novel material with simultaneous detection towards CO, H_2 and CH_4. Sensors and Actuators B: Chemical 66(1): 225–27. doi:https://doi.org/10.1016/S0925-4005(00)00371-3.

Majumdar, Sanhita. 2015. The effects of crystallite size, surface area and morphology on the sensing properties of nanocrystalline SnO_2 based system. Ceramics International 41(10, Part B): 14350–14358. doi:https://doi.org/10.1016/j.ceramint.2015.07.068.

Meng, Zheng, Robert M. Stolz, Lukasz Mendecki and Katherine A. Mirica 2019. Electrically-transduced chemical sensors based on two-dimensional nanomaterials. Chemical Reviews 119(1): 478–598. American Chemical Society. doi:10.1021/acs.chemrev.8b00311.

Okeke, Izunna S., Kenneth K. Agwu, Augustine A. Ubachukwu and Fabian I. Ezema. 2022. Influence of transition metal doping on physiochemical and antibacterial properties of znonanoparticles: a review. Applied Surface Science Advances 8: 100227. doi:https://doi.org/10.1016/j.apsadv.2022.100227.

Paolesse, Roberto, Sara Nardis, Donato Monti, Manuela Stefanelli and Corrado Di Natale. 2017. Porphyrinoids for chemical sensor applications, 2517–2583. doi:10.1021/acs.chemrev.6b00361.

Pilliadugula, Rekha and N. Gopalakrishnan. 2022. Morphology dependent room temperature CO_2 sensing of β-Ga_2O_3. Materials Today: Proceedings. doi:https://doi.org/10.1016/j.matpr.2022.02.366.

Prakash, Jai, Samriti Ajay Kumar, Hongliu Dai, Bruno C. Janegitz, Venkata Krishnan, Hendrik C. Swart and Shuhui Sun. 2021. Novel rare earth metal–doped one-dimensional TiO_2 nanostructures: fundamentals and multifunctional applications. Materials Today Sustainability 13: 100066. doi:https://doi.org/10.1016/j.mtsust.2021.100066.

Rao, C.N.R., G.U. Kulkarni, P. John Thomas and Peter P. Edwards. 2002. Size-dependent chemistry: properties of nanocrystals. Chemistry–A European Journal 8(1): 28–35. John Wiley & Sons. doi:https://doi.org/10.1002/1521-3765(20020104)8:1<28::AID-CHEM28>3.0.CO;2-B.

Singh, Gurpreet, Virpal and Ravi Chand Singh. 2019. Highly sensitive gas sensor based on er-doped SnO_2 nanostructures and its temperature dependent selectivity towards hydrogen and ethanol. Sensors and Actuators B: Chemical 282: 373–383. doi:https://doi.org/10.1016/j.snb.2018.11.086.

Tischner, Alexandra, Thomas Maier, Christoph Stepper and Anton Köck. 2008. Ultrathin SnO_2 gas sensors fabricated by spray pyrolysis for the detection of humidity and carbon monoxide. Sensors and Actuators B: Chemical 134(2): 796–802. doi:https://doi.org/10.1016/j.snb.2008.06.032.

Wan, Xuejuan, Jilei Wang, Lianfeng Zhu and Jiaoning Tang. 2014. Gas sensing properties of Cu_2O and its particle size and morphology-dependent gas-detection sensitivity. Journal of Materials Chemistry A 2(33): 13641–13647. The Royal Society of Chemistry. doi:10.1039/C4TA02659D.

Wang, Xu-dong and Otto S. Wolfbeis. 2016. Fiber-optic chemical sensors and biosensors (2013–2015). Analytical Chemistry 88(1): 203–227. American Chemical Society. doi:10.1021/acs.analchem.5b04298.

Wang, Xu-dong and Otto S. Wolfbeis. 2020. Fiber-optic chemical sensors and biosensors (2015–2019). Analytical Chemistry 92(1): 397–430. American Chemical Society. doi:10.1021/acs.analchem.9b04708.

Xu, Chaonan, Jun Tamaki, Norio Miura and Noboru Yamazoe. 1991. Grain size effects on gas sensitivity of porous SnO_2-based elements. Sensors and Actuators B: Chemical 3(2): 147–155. doi:https://doi.org/10.1016/0925-4005(91)80207-Z.

Yamazoe, N., Y. Kurokawa and T. Seiyama. 1983. Effects of additives on semiconductor gas sensors. Sensors and Actuators 4: 283–289. doi:https://doi.org/10.1016/0250-6874(83)85034-3.

Yao, Ming-Shui, Wen-Hua Li and Gang Xu. 2021. Metal–organic frameworks and their derivatives for electrically-transduced gas sensors. Coordination Chemistry Reviews 426: 213479. doi:https://doi.org/10.1016/j.ccr.2020.213479.

Yoon, Yeosang, Phuoc Loc Truong, Daeho Lee and Seung Hwan Ko. 2021. Metal-oxide nanomaterials synthesis and applications in flexible and wearable sensors. ACS Nanoscience Au 2(2): 64–92. American Chemical Society. doi:10.1021/acsnanoscienceau.1c00029.

Zairi, S., C. Martelet, N. Jaffrezic-Renault, R. M'gaïeth, H. Maâref and R. Lamartine. 2001. Porous silicon a transducer material for a high-sensitive (Bio) chemical sensor: effect of a porosity, pores morphologies and a large surface area on a sensitivity. Thin Solid Films 383(1): 325–327. doi:https://doi.org/10.1016/S0040-6090(00)01607-2.

Zhang, Chao, Guifang Liu, Xin Geng, Kaidi Wu and Marc Debliquy. 2020. Metal oxide semiconductors with highly concentrated oxygen vacancies for gas sensing materials: a review. Sensors and Actuators A: Physical 309: 112026. doi:https://doi.org/10.1016/j.sna.2020.112026.

Chapter 2

Measurement Techniques for Chemical Sensing

2.1 INTRODUCTION

Molecular recognition is very important for environmental and industrial applications. For this purpose, various analytical techniques have been developed which are able to detect physical parameters like pressure or temperature and obtain the information about chemical compounds. These analytical techniques for example gas chromatography–mass spectrometry (GC/MS), mass spectrometry (MS), and atomic absorption spectroscopy (AAS) have been widely used for the quantification of chemical compounds with great accuracy and stability. However, these systems are highly expensive, time-consuming, and not handy, and therefore, limiting their use for on-site monitoring applications (Cao et al. 2021; Hasan et al. 2020). Thus, there is a need to develop a simple, robust, low-cost, sensitive, and reliable sensory platform for chemical and biochemical detection. These advanced sensory platforms are able to detect a minute change in the surrounding environment. These are highly dependent upon the materials' physical and chemical properties.

Now, a new era has begun in the development of chemical sensors for many diverse applications. Special focus is being paid to design and develop a simple, miniaturized, portable, easy to operate, sensitive, selective, reliable, and low-cost chemical sensor (Koo et al. 2019; Lee et al. 2017; Meng et al. 2019; O'Sullivan and Guilbault 1999; Tsouti et al. 2011; Wang and Wolfbeis 2016). These sensors are very simple and provide a wireless online monitoring feature, and therefore, highly in demand for environmental and industrial applications.

In this chapter, a few widely adopted and reliable chemical sensing platforms are reviewed. Their working mechanism and importance in chemical sensing are also presented.

2.2 OPTICAL APPROACH

An optical sensor can be realized via a light-matter interaction. An electromagnetic (EM) spectrum is the source of this type of interaction. An EM spectrum consists of different energy bands which possess specific wavelengths. If the frequency increases, the corresponding wavelength decreases and thus, increases the radiation energy. The spectrum is divided into various categories starting from (with the increase in wavelength) NMR, ESR, microwaves, infrared, visible-ultraviolet, X-ray, and gamma ray, respectively (Houck and Siegel 2015; Jespersen 2006). To make a chemical sensor in a specific region, it depends upon the interaction of radiation and the material. Mostly, to fabricate a chemical sensor, certain spectroscopy like UV-vis and infrared are extensively used. This is because radiation with lower energy only changes the rotation of the molecules whereas radiation with higher energy can excite the molecule into a higher energy state and thus, not be feasible for the effective chemical sensing applications. A high thermal energy is required to vaporize or atomize a sample in order to perform the atomic spectroscopy and hence, hard to utilize for the development of a chemical sensor (Beć et al. 2020; House 2018; Jespersen 2006; Lv et al. 2022). Actually, not all the spectroscopic measurements are useful for the development of a chemical sensor due to certain limitations related to their radiation energy, wavelengths, and detected frequency.

2.2.1 Interaction of Light with Matter

When radiation interacts with material or analyte, mostly it interacts through the following phenomenon.

1. Reflection
2. Transmission
3. Refraction
4. Scattering

For reflection, we need to consider the refractive index of the two mediums, i.e., one through which the radiation travels and other one for which radiation to interact. So, by Snell's law, the angle of incidence is related to the angle of reflection. Hence, for two media have different refractive indices of n_1 (less dense than n_2) and n_2 (denser than n_1) the law is given as follows (Gilbert 2022).

$$n_1 \sin(\theta_1) = n_2 \sin(\theta_2) \tag{2.1}$$

Thus, to achieve the reflection, the angle of incidence should attain the critical angle. The critical angle is the angle at which the incident light travels along the surface and makes 90 degrees with the normal. When the angle of incidence is more than the critical angle, then light travels inside the second medium. The incident light then travels further by the total internal reflection principle. The TIR principle is commonly used in fiber optics-based sensors. Both reflection and refraction are used for the fabrication of a chemical sensor. The refracted ray also lies on the plane and on the other side of normal. When an angle of incidence is higher than critical angle, the radiation is said to be refracted.

Generally, the reflection from an optically opaque surface can be specular or mirror-like. Then all the reflected rays are in phase with one another. While diffused radiation can emerge when the surface is not flat. The radiation undergoes multiple scatterings and partial absorption within the medium. The scattered radiations are not in phase with one another and they get randomly scattered from the surface. However, these kinds of radiations do have chemical information about the material sample and can be used in material analysis. The reflected intensity of the diffused radiation mainly depends upon the composition of the material, and therefore, the adsorbed concentration inside the material is given by the Kubelka-Munk equation (Militký 2011; Shibuya and Kawase 2013):

$$F = \frac{(1 - R)^2}{2R} = \frac{\varepsilon C}{S} \tag{2.2}$$

where, $F(R)$ is the Kubelka-Munk function, R is reflectance, ε is the molar absorptivity, C is concentration, and S is the scattering coefficient.

Depending upon the sensing principle and geometrical arrangements, different types of optical approaches have been implemented for detection of various chemicals. The information of these techniques is described as follows:

2.2.2 Absorbance-based Sensors

Absorbance mechanism is widely used for chemical detection. Generally, an EM radiation with high energy (means lower wavelength) once absorbed by the sample will cause an excitation of ground state electron to a higher energy state. The absorbance occurs when the photon energy matches the difference between two energy levels E_2 and E_1. More clearly, at some photon energy, the electron transition occurs from the bonding level (π)

to an anti-bonding level (π^*) or it may occur between non-bonding level (n) to an anti-bonding level (π^*) causing absorbance (Gilbert 2022; Houck and Siegel 2015; Wypych 2015). For example, in case of benzene, the absorbance occurs due to the move from bonding state (π) to the anti-bonding state (π^*) at a photon energy of 4.88 eV at an absorbed wavelength of 254 nm.

This is the most common detection technique relying on the Beer-Lambert law. Lambert proposed that the intensity of the EM radiation (monochromatic) is the logarithmic function of increase in length of EM radiation path. Once the EM radiation interacts with the analyte, the intensity of EM radiation starts decreasing. Beer stated that the transmission of EM radiation in a liquid solution of analyte is exponential function of concentration of analyte. Thus, both statements combined and generated a unique law named as Beer-Lambert law (Wypych 2015).

Therefore, the absorbance is given by following expression:

$$A = -\log(T) = -\log\left(\frac{I}{I_0}\right) = \log\left(\frac{I_0}{I}\right) \tag{2.3}$$

where, A is absorbance, T is transmittance, I_0 is source intensity and I is final intensity detected at the detector.

Thus, for a solute with concentration c (molecule cm^{-3}), A is given by Beer-Lambert law as follows:

$$A = \sigma l c \tag{2.4}$$

where, σ is the absorption cross-section in cm^{-2}-molecule^{-1} and l is optical length of the gas cell in cm, respectively. However, this is mainly used for low concentration of analyte and monochromatic source. Thus, at higher concentration of analyte and for a variable lengths of the cell, the Beer-Lamberts law is modified as per the following:

$$I = I_0 e^{-\sigma l c} \tag{2.5}$$

$$\frac{dI}{dl} = -\sigma c I_0 e^{-\sigma l c} \quad \text{and} \quad \frac{dI}{dc} = -\sigma l I_0 e^{-\sigma l c} \tag{2.6}$$

The absorption spectrum of molecules in solution exhibits much broader band spectra than the gas phase. This occurs due to the proximity of vibrational levels and hence, possesses large number of the interactions in solution state (shown bottom). However, in case of gas or vapor, a sharp spectrum is obtained. Therefore, an EM wave should get absorbed in the solution phase or in gas phase to generate an absorption spectrum.

Non-linearity can also be generated in the absorption spectrum. If the monochromatic source is used instead of polychromatic light, then very narrow absorption spectra could be obtained. Generally, in case of absorption, if the spectral width is smaller than the full-width-half maximum (FWHM) of absorption peak then maximum light intensity gets absorbed.

2.2.3 Fluorescence and Phosphorescence

Nowadays, fluorescence-based technique is gathering particular interest for researchers for various chemical sensing due to their simplicity and wide detection approach. Various possible schemes such as intensity-based, phase modulation, and time intensity ratio have been explored to detect the number of analytes with a good accuracy and high sensitivity.

The fluorescence and phosphorescence are very important features of luminescence. These come under the cold emission category. In scientific language, these are the non-equilibrium phenomenon that occur above the thermal energy of the body in which the duration of the emission is beyond the period of the light oscillations. The fluorescence emission occurs from singlet state and the typical life time of the emitted radiation varies from 10^{-10} s to 10^{-7} s. Whereas this phenomenon is relatively slow in case of phosphorescence and it occurs from triplet states and its normal life-time is in the range of 10^{-5} s to $> 10^{+3}$ s (Fleming 2017). In fluorescence, the de-excitation occurs via radiative decay process in the electronic excited states of the molecule. Generally, the fluorescence is observed in lowermost vibrational level of the excited electronic state to the electronic ground state. Also, most of the fluorescence spectra shift towards lower energy than that of absorption spectrum. Therefore, the smallest wavelength in fluorescence is the lengthiest wavelength in absorption spectrum.

Phosphorescence is the radiative transition that occurs from triplet state (T_1) to ground state (S_0). Generally, phosphorescence spectra are observed at a higher wavelength side than fluorescence because the lowest vibrational level of triplet state T_1 is lower than the singlet state S_1. Generally, the luminescent intensity in linear relationship with concentration (c) and optical path-length (l) is given by:

$$I_L = \phi_L I_0 k \varepsilon_\lambda l_c \tag{2.7}$$

where, ϕ_L is the luminescence quantum yield, I_0 is the intensity of the excitation light, κ is the instruments' related parameter such as geometrical parameter of the sample, source and collector related collection efficiency, and ε_λ is the luminophore absorption coefficient at the excitation wavelength in the unit of $dm^3mol^{-1}cm^{-1}$.

According to the Stern-Volmer equation, the ratio of fluorescence intensity in the absence (I_0) and in presence (I) of the quencher is given by (Devi et al. 2018; Gehlen 2020):

$$\frac{I_0}{I} = 1 + K_{SV}[Q] \tag{2.8}$$

where, K_{SV} is the 'Stern-Volmer constant' and Q is the quencher concentration.

However, for fluorescence-based sensors, the fluorescence intensity may fluctuate due to instrumental parameters such as aging of light source and detector, effect of ambient light, and fluorophore leaching and photodegradation process. Therefore, to remove these drawbacks, the luminescence lifetime (τ) needs to be considered.

Thus, the Stern-Volmer equation is modified as follows:

$$\frac{\tau_0}{\tau} = 1 + K_{SV}[Q] \tag{2.9}$$

One of the advantages of the above equation is to generate the linear response of fluorescence-based sensing devices.

2.2.4 Optical Fiber-based Approach

An optical fiber sensor (OFS) approach is an excellent choice in chemical sensing applications due to its peculiar properties as that of the electrical sensors. If any external stimuli changes, any one of the properties of OFS could change; such as: (1) polarization (2) intensity (3) phase (4) wavelength and (5) spectral distribution. Change in either one of these parameters proves to be the fundamental basis for the implementation of various optical fiber chemical sensors (Pawar and Kale 2019; Wang and Wolfbeis 2016).

2.2.4.1 Basics of fiber optics

2.2.4.1.1 *Operating principle of the fiber*

Optical fiber acts as a waveguide that propagates the light from initial launching point to another release point. The light propagates inside the optical fiber by a 'total internal reflection' phenomenon. According to Snell's law, if the incident angle is greater than the critical angle, the total internal reflection occurs (Gilbert 2022).

2.2.4.1.2 *Structure of the optical fiber*

Optical fiber consists of three parts namely, core, cladding, and coating, respectively.

1. The core part is dielectric in nature and has a slightly higher refractive index than cladding part. The light is guided through the core part.

2. The cladding part is also dielectric in nature and it prevents the light to escape in air. It has a slightly lower refractive index than the core.

3. The buffer coating is made up of polymeric material. This is to protect the fiber from the environment or any physical damage. It gives strength to the fiber.

The refractive indices of the core and cladding of the fiber can be easily altered by adding the compatible oxide material in pure silica. The oxides such as TiO_2, GeO_2, P_2O_5, Al_2O_3 increase the refractive index while B_2O_3, and Fluorine lower the refractive index.

2.2.4.2 Fiber optic interferometric techniques

An interferometer is an optical device that generates an interference pattern through the superposition of two or more waves. Interferometers are divided into two categories viz: wave-front division and amplitude division interferometers. In case of optical fiber sensing domain, there exist four main types of optical fiber interferometric techniques (Figure 2.1), namely Michelson, Mach-Zehnder, Sagnac, and the Fabry-Perot interferometer. These techniques are well-known for their varied geometry, working principle and sensitivity (Elosua et al. 2017; Pawar and Kale 2019; Urrutia et al. 2015). The Michelson, Mach-Zehnder, and Sagnac interferometers work on two-beam interference interferometer and Fabry-Perot interferometer works on multi-beam interference interferometric principle.

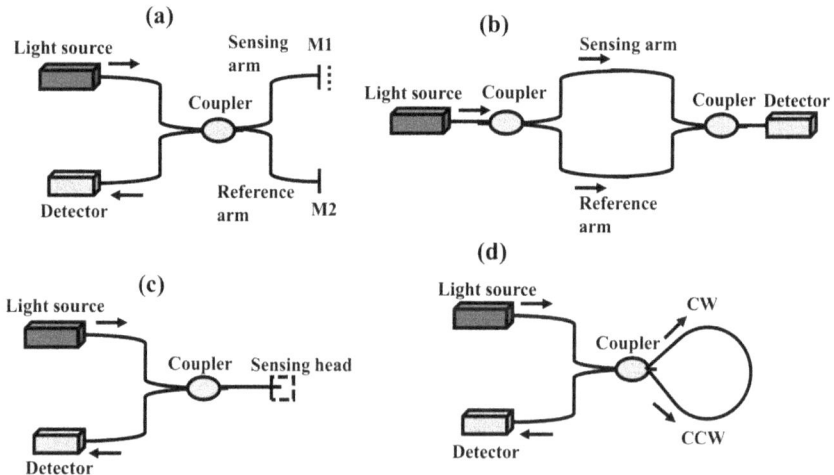

Figure 2.1 Basic configuration of different optical fiber interferometric techniques. (a) Michelson interferometer, (b) Mach-Zehnder interferometer, (c) Fabry-Perot interferometer, and (d) Sagnac interferometer.

2.2.4.2.1 Michelson Interferometer (MI)

In this interferometer, a beam coupler acts as a splitter and a re-combiner. If the two-arm length is the same in exact perpendicular position, then only one output is generated and other output goes to the light source (Shown in Figure 2.1a). If the arm length is not same, it generates two outputs, which are accessible. This interferometer operates in reflection mode. Among the two arms, one turns as a sensing arm and other is referred to as a reference arm. If the path length is non-zero, in between two arms, either a constructive or a destructive interference will occur (Pawar and Kale 2019; Urrutia et al. 2015).

2.2.4.2.2 Mach-Zehnder Interferometer (MZI)

The MZI consists of two arms called a sensing arm and a reference arm. The incident light splits into two beams at one beam coupler and combines at another beam coupler. The interference generated at the second beam coupler depends on the optical path difference (OPD) between two arms as shown in Figure 2.1b. The output is calibrated with respect to the reference arm. In recent times, in-line types of MZIs have been reported to be very effective for various applications. Therefore, the physical length in both arms is same, but the optical path lengths could be different because of the modal dispersion. This occurs when refractive index of one of fiber components (core or clad) can be manipulated using the sensing parameter (Pawar et al. 2016; Urrutia et al. 2015).

As shown in Figure 2.1b, the principal mechanism of the MZI is expressed in terms of two-beam optical interference equation. The total intensity of interference pattern in terms of core and cladding modes is given by the equation (Pawar and Kale 2019).

$$I = I_{\text{core}} + I_{\text{clad}} + 2\sqrt{I_{\text{core}}I_{\text{clad}}}\ \cos(\phi) \qquad (2.10)$$

where, I_{core}, I_{clad}, are the core and cladding mode intensities; and ϕ is the phase difference.

The principle of an inline MZI is explained in the following manner. Generally, the core light gets partially bent at the first splicing point. The diffracted light travels through the cladding part of the fiber called as cladding modes and core mode light continues to travel in the core region of the fiber. Thus, there is an opportunity for cladding modes to interact with the surrounding medium. These cladding modes travel nearby at clad-air boundary. These two core and cladding modes are recombined at the second collapsing points. Once they interfere with each other, an interference pattern is formed. The interference is produced owing to the variance in propagation constant among the core and cladding modes of the fiber (Pawar and Kale 2016; Urrutia et al. 2015).

2.2.4.2.3 Fabry-Perot Interferometer (FPI)

FPI contains two parallel reflecting mirrors separated by some distance and thus, form an etalon. Typical optical fiber based FPI is shown in Figure 2.1c. When a light ray enters the FP cavity, it gets multiplied internally and gets reflected back and forth among the mirrors. The interference is produced by superimposing the transmitted and reflected beams. For optical fiber, FP cavity might be located internally or externally and named intrinsic FPI and extrinsic FPI, respectively (Pawar and Kale 2019; Ugale et al. 2016; Urrutia et al. 2015).

The FPI forms in between two reflecting mirrors. For optical fiber case, the thin film is deposited at the end of the fiber tip forming a Fabry-Perot cavity. The cavity consists of two interfaces such as fiber-film (R_1); and film-air (R_2), with corresponding reflection coefficients as $R_1 = r_{12}$ and, $R_2 = r_{23}$ respectively.

By considering Fresnel's reflectance coefficients at the cavity surfaces, the total reflectance of FP cavity is dependent upon numerous factors like refractive indices, incident angle, and the polarization of the incident wave. The total reflectance of the FPI can be expressed as (Pawar and Kale 2019; Pawar et al. 2018):

$$R_{123} = |r|^2 = \frac{r_{12}^2 + r_{23}^3 + 2r_{12}r_{23}\cos 2\phi}{1 + r_{12}^2 r_{23}^3 2r_{12}r_{23}\cos 2\phi} \tag{2.11}$$

The reflectivity coefficients of FP cavity in between, i.e., first reflecting surface mentioned as fiber-film (r_{12}) and second reflecting surface mentioned as film-air (r_{23}) interfaces are given by following manner (Pawar et al. 2017):

$$r_{12} = \frac{n_{eff} - n_{film}}{n_{eff} + n_{film}}; \quad r_{23} = \frac{n_{film} - n_{ext}}{n_{film} + n_{ext}}; \quad \phi = \left(\frac{4\pi}{\lambda}\right)n_{film}d \tag{2.12}$$

where, n_{eff} is the effective refractive index of optical fiber; n_{ext} is the refractive index of external medium (external stimuli) (in our case, it is gas concentration); and ϕ is the propagation phase in the coated film.

2.2.4.2.4 Sagnac Interferometer (SI)

This kind of interferometer contains a circular loop formed by splitting an input light beam at the beam coupler. One beam propagates in a clockwise direction and another beam propagates in the counter clockwise direction. Each beam possesses different polarization states. These two beams again combine at the beam coupler, generating an interference pattern. The SI is shown in Figure 2.1d. The OPD is calculated by the polarization dependent propagating velocity of the mode guided alongside the loop (Urrutia et al. 2015).

OFSs offer a lot of advantages over electrical based sensors. The benefits are listed below (Elosua et al. 2017; Pawar and Kale 2019; Urrutia et al. 2015; Wang and Wolfbeis 2016).

1. They are non-electrical.
2. They are immune to electromagnetic interference (EMI) and radiofrequency interference.
3. They can be easily multiplexed and hence, good for distributed sensing.
4. They are lightweight, robust and resistant to hostile environments.
5. They are flexible and data transmission is secure.
6. They are highly sensitive, and have the capability of multi-parameter sensing.

2.3   CHEMIRESISTIVE APPROACH

The chemiresistive-based sensors sensing mechanism (Figure 2.2) is dependent upon a change in the resistance of the material and it is decided primarily by the reaction occurs between the analyte and adsorbed oxygen species on the surface of the material. However, materials' other properties (as discussed in Chapter 1) also play a crucial role in deciding the sensor performance. With respect to the variation in the analyte concentration, the depletion layer or the potential barrier height will get altered and hence, the conductance of the material is changed. This method is very cheap, simple to fabricate, offers a great interface with the electronic data convertors, and good flexibility, etc. (Das et al. 2022; Koo et al. 2019; Majhi et al. 2021).

The most common chemiresistive system based on an indirect-heated type derived from the Figaro-Taguchi type (Figure 2.2a, b) (TGS) sensor and microelectromechanical system (MEMS) type sensor (Figure 2.2c, d) is presented.

Ideally, neither too many nor too less electrons are useful in gas sensing. Therefore, the most trustable chemiresistive-based materials are metal oxides (MOXs) due to their unique structural design which offer the right balance between availability of free charges for reaction, their limited concentration, low-cost synthesis, and good stability. For too many electrons structure, e.g., metal, the gain of the reaction does not affect the analyte interaction, and therefore, the MOXs are the best for such detections. The semiconductor material band-gap is small enough at elevated temperature and thus, electrons will be available for the interaction with analyte (Cao et al. 2020, 2021; Das et al. 2022). Some of the most widely used MOXs are SnO_2, ZnO, and CuO, etc. In case of gas sensing, when reducing gas (like CO, H_2S) interacts with the n-type

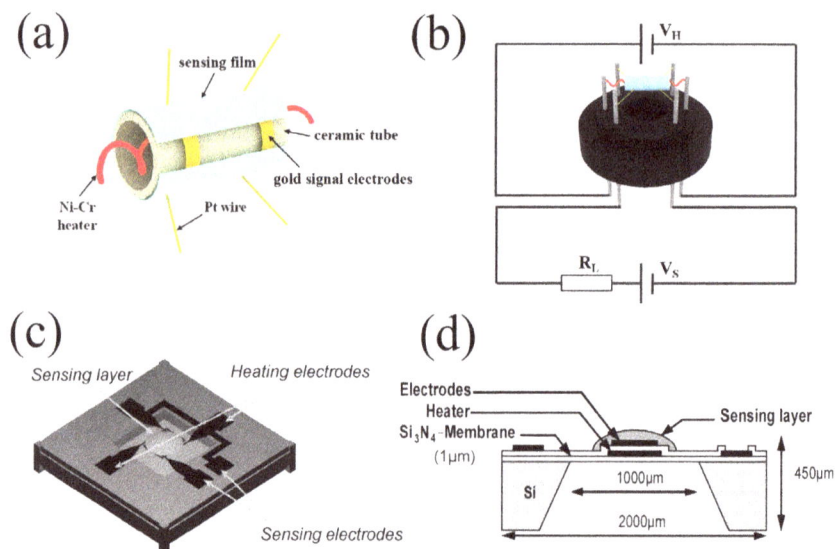

Figure 2.2 Typical chemiresistive sensing elements. Figaro-Taguchi type (TGS) sensor: (a) Indirect-heated type sensor, (b) Its equivalent test circuit; Microelectromechanical system (MEMS) type: (c) 3D view and side view of the device, and (d) Its complete view along the dimensions. [Reproduced with permission (Yao et al. 2021) Copyright 2021, Elsevier].

MOX, the conductivity gets increased due to the withdrawal of electrons from the conduction band. Whereas, a decrease in conductivity is observed for p-type MOX due to donation of electrons to the conduction band (Das et al. 2022; Korotcenkov 2007). Generally, three types of oxygen species (O_2^-, O^-, O^{2-}) are observed depending on the working temperature. The presence of oxygen molecules plays a significant role in chemiresistive-based sensing. The oxygen vacancies in the form of molecular–ionic species at temperature below 150 °C get converted into atomic–ionic species at a higher temperature (Cao et al. 2020, 2021; Das et al. 2022).

At an appropriate working temperature, the adsorbed oxygen species create a depletion layer surrounding the material. In this case, depending upon the type of gas, i.e., either reducing or oxidizing, the transfer of the electrons from the gas to the material and vice versa occurs. When gas interacts with the material, it changes the depletion width or potential barrier and thus, the resistance of the material. Numerous composite materials have been used for the chemical sensing application (Cao et al. 2021; Das et al. 2022; Koo et al. 2019; Korotcenkov 2007). The typical energy band diagram for different composite formations namely n-n, p-p, and n-p is depicted in Figure 2.3.

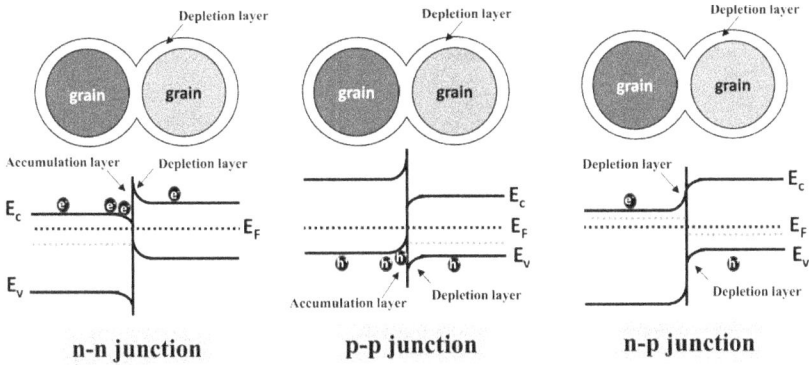

Figure 2.3 The energy band diagram upon interaction with different conductivity metal oxides materials.

2.4 ELECTROCHEMICAL APPROACH

There are many electrochemical techniques that are reported for the detection of numerous chemical and biological species. The most common techniques have been discussed below. In electrochemical measurements (Figure 2.4), the following terms are commonly used (Zhang and Hoshino 2019).

1. **Electrochemical cell:** Generally, the electrochemical measurement contains two half-cells. In the half-cell, one electrode is located in a salt solution which leads to charge separation at the metal-electrode interface (Figure 2.4a). The placing of two half-cells aside makes a complete electrochemical cell and these two cells are connected by a salt bridge.

2. **Daniell cell:** It is a galvanic cell consisting of Zn and Cu electrodes and these are dipped into the copper (II) sulfate and zinc sulfate solutions, which forms a reduction and oxidation reaction. The copper acts as an oxidant and zinc works as a reductant. The complete Daniell cell reaction is expressed as following:

$$Cu^{2+}(aq) + Zn(s)Cu(s) + Zn^{2+}(aq) \qquad (2.13)$$

3. **Electrode potential:** Due to the use of two metals in an electrochemical cell, the current flow is dependent upon the ionization of the metals, and therefore, a reference potential is used to quantify them via calculating the comparative potentials of the metals. Generally, the silver-silver chloride (Ag/AgCl) electrode and the saturated calomel electrode (SCE) are commonly used as reference electrodes.

4. **Nernst equation:** The Nernst equation is related to the electrode potential temperature and the standard-state electrode potential. The Nernst equation can be expressed as:

$$E_{\text{cell}} = \frac{RT}{nF} \ln \frac{[A]_1}{[A]_2} \qquad (2.14)$$

where, F is the Faraday constant (96, 485 C/mol), T is the temperature and A is related to the effective thermodynamic concentration of the oxidant and the reductant, respectively.

Among the various electrochemical techniques, a few techniques that are commonly used are voltammetry, potentiometry, amperometry and conductometric. Here, only voltammetry and amperometry are discussed. The voltammetry consists of various methods such as cyclic voltammetry, square wave voltammetry, differential pulse voltammetry, etc.

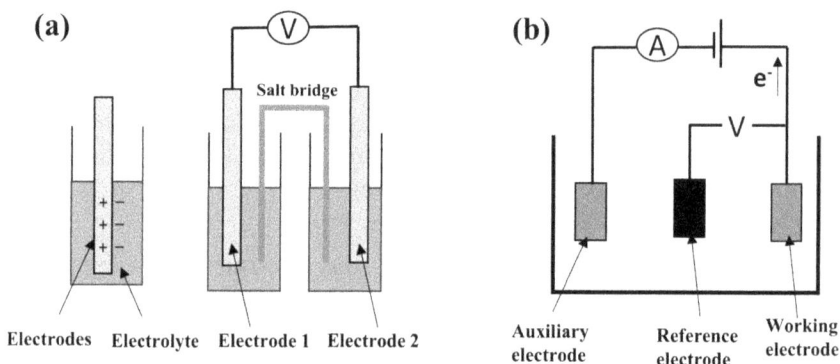

Figure 2.4   (a) Electrochemical cell consisting of a half cell and combination of two half-cells, (b) Setup of voltammetry.

2.4.1   Voltammetry

This technique is widely used in electrochemical measurements and it gives qualitative information regarding the properties and characteristics of the electrochemical processes, thermodynamics and kinetics of redox reactions. In this method, the analyte information is extracted by varying potential and measures the corresponding changes in the current. Generally, it contains basic electrodes such as working, auxiliary, and reference electrodes (Figure 2.4b). The reference electrode plays a crucial role in controlling and steadying the potential of working electrode. The accuracy can be improved by using three electrode system. As the potential is applied across the analyte, the level of electron transmission in the electrochemical reactions can be measured. The typical signal spectrums are shown in Figure 2.5. The

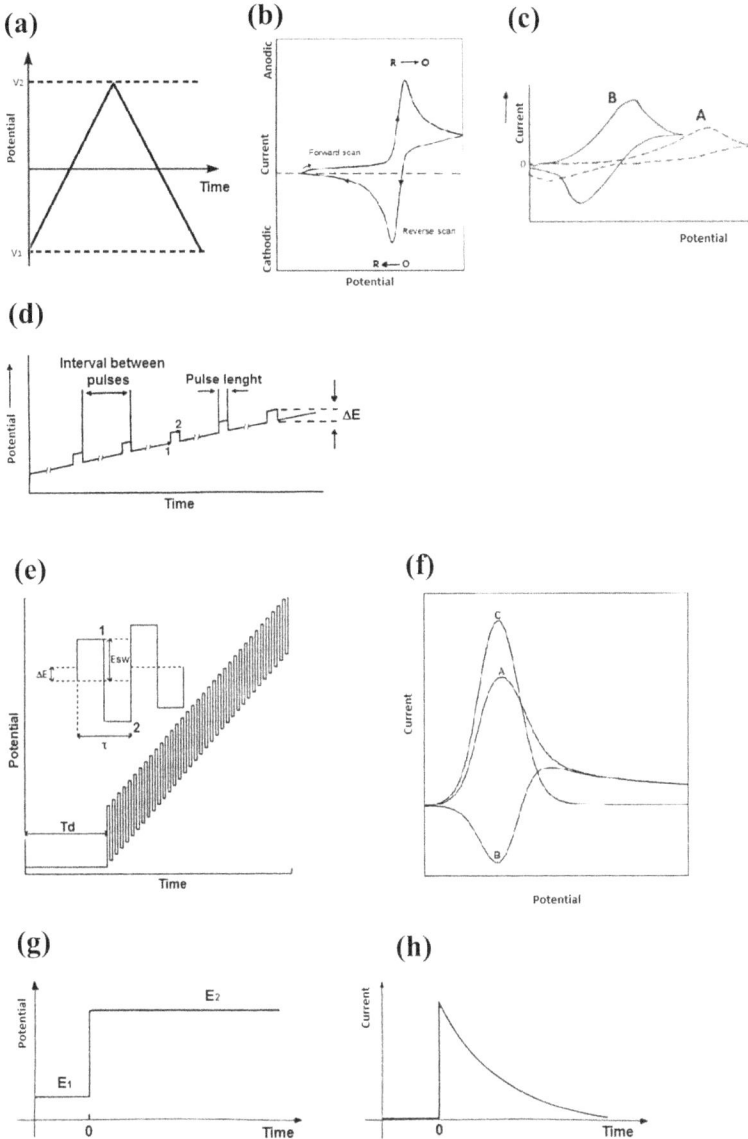

Figure 2.5 (a) Potential-time excitation signal in cyclic voltammetry. (b) Typical cyclic voltammogram for a reversible redox process (O + ne R). (c) Voltammogram for Quasi-reversible system (Curve B) and irreversible system (Curve A). (d) Potential-time excitation signal in differential pulse voltammetry. (e) Potential-time excitation signal in square wave voltammetry. (f) Square wave voltammogram for a reversible process. Curve A: Forward current. Curve B: Reverse current. Curve C: net current. (g) Potential-time waveform in a chronoamperometric experiment. (h) Current-time response in a chronoamperometric experiment. [Reprinted from Publication Ref. (Hoyos-Arbeláez et al. 2017); copyright 2017, Elsevier].

current produced in the electrochemical reaction rises with increasing voltage to potential of analyte electrochemical reduction. When the voltage comes to its original value, the current upsurges in the reverse polarity and reduces. Thus, the reverse scan also gives the reaction reversibility. The concentration of catalysts also plays an important role in deciding the shape of the voltagram (Hoyos-Arbeláez et al. 2017).

2.4.2 Amperometry

The amperometry sensor measures the change in current that is caused by the migrating ions which create the potential difference due to applied voltage. The semipermeable membrane allows the analyte species into the electrolyte solution. The potential difference between the two electrodes is applied to it. The reaction at the working electrode is given as following (Hoyos-Arbeláez et al. 2017; Patel 2020; Zhang and Hoshino 2019):

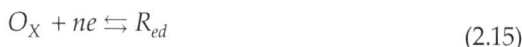

$$O_X + ne \leftrightarrows R_{ed} \tag{2.15}$$

In this process, the species are oxidized at anode and become positive charged cation by losing the electrode. Those electrons are consumed by an anode where other species are reduced and become negatively charged anions by gaining the electrons. Therefore, this generates a diffusion-controlled current, which is given by following equation:

$$I = \frac{nFAD(C_{\text{bulk}} - C_{x=0})}{L} \tag{2.16}$$

where, n is number of electron transfer, F is Faraday number (96,480 C/mol), A is surface area of the electrode, D is analyte diffusion coefficients, L is thickness of diffusion layer, and C_{bulk} and $C_{x=0}$: concentration of analyte in the bulk and concentration at the surface of the electrode, respectively.

It is concluded that the current generated is directly proportional to the number of electrons generated and hence, the species get reduced. At no species at anode, the weak zero current flowing will through it. When the analyte is passed through the permeable membrane, it gets reduced at the cathode and forms a plateau. This also represents an equilibrium, which signifies that the current generated is straightforwardly proportional to the concentration of the test sample.

From a device point of view, the sensitivity of the electrode is first measured by doing calibration as following:

$$S = \frac{I_{\text{calibration}}}{C_m} \tag{2.17}$$

where, I is current in the calibration, C_x is the known concentration of the substance m in the calibration sample.

Thus, the concentration of the substance m in the sample is given by:

$$C_m(\text{sample}) = \frac{I_{\text{sample}} - I_0}{S} \qquad (2.18)$$

where, I_{sample} is the maximum sample current, I_0 is the zero current when no sample, and S is the sensitivity of the electrode obtained from the calibration measurement.

2.5 CAPACITIVE APPROACH

A capacitive type of sensing consists of a capacitor in which the two parallel conductive plates with area S (in m^2) distanced by a finite distance d (in m) form a capacitor (Figure 2.6). The dielectric material placed between the two parallel plates decides the storage capacity of the device. Overall, the capacitance C of the system is dependent upon the dielectric constant of free space ε_0 (in F/m) and dielectric material ε_r (in F/m), surface area of the parallel plates, and distance. The capacitance of the materials can easily change by changing the distance between the plates (Zhang and Hoshino 2019).

The C is the function of the three parameters and is given by the following:

$$C = f(\varepsilon, S, d) \qquad (2.19)$$

$$C = \frac{\varepsilon S}{d} = \frac{\varepsilon_0 \varepsilon_r S}{d} \qquad (2.20)$$

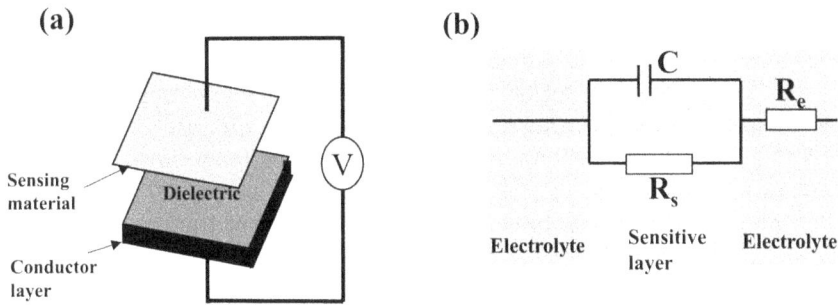

Figure 2.6 (a) Schematic of a capacitive-based sensor, (b) Equivalent electric circuit diagram.

When such kind of sensor is exposed to an analyte, the corresponding dielectric constant of the material will get changed and hence, the corresponding change in the capacitance will be observed.

If we differentiate equation 1 with respect to distance then we will get the sensitivity of the device.

$$K = \frac{\Delta C}{\Delta d} = -\frac{\varepsilon_0 \varepsilon_r S}{d^2} \qquad (2.21)$$

Therefore, to achieve the highest sensitivity, one needs to consider a smaller separation between the plates of the capacitor.

The capacitive sensor response can be measured as the following equation:

$$R = \frac{C_T(\text{analyte maximum concentration}) - C_0(\text{analyte at zero concentration})}{C_0(\text{analyte at zero concentration})}$$

$$(2.22)$$

where, C_T is the capacitance change under analyte concentration and C_0 is the sensor's normal state capacitance.

When any analyte (non-polymeric or polymeric film) gets adsorbed on the material, which may induce alteration in the dielectric constant. The impedance of the sensor can be found by putting this sensor as an *RC* circuit (Kummer et al. 2004; Panasyuk et al. 1999).

Commonly used capacitive sensors in chemical sensing are divided into two types such as:

1. Interdigitated electrodes (IDEs) and 2. Electrode–solution interfaces.

In IDEs-based sensors, the change in capacitance is monitored in presence of interaction between immobilized probes and target molecules. In this case, the electrodes may be thicker and their edge effects are neglected;

For an electrode–solution boundary, the capacitance depends upon the dielectric constant and more importantly on the rise of its width and the movements of ions and water molecules from surface (Tsouti et al. 2011).

Some of the key points to be considered are that the receptor layer should have low leakages and low thickness for effective adsorption. The change in the dielectric constant leads to alter the capacitance value of the device. However, this is not useful for gas detection as the relative permittivity of maximum inorganic gases (excluding H_2O) is alike, and therefore, needs to be considered for the dielectric thickness. Sometimes, an electrode area will also alter in case of humidity sensing. At present, varying the distance (d) between the plates is considered a different way in biosensing field. The IDE method is expensive and requires expensive fabrication techniques such as lithography, but is miniaturized and offers high sensitivity. While the electron-solution interfaces method is low-cost, rapid, and label-free.

2.6 FIELD EFFECT TRANSISTOR APPROACH

At present, the field-effect transistors (FETs) are attracting a great deal of attention in chemical sensing because of their simplicity, high sensitivity, fast detection, flexibility, and stable performance. The FET measures the conductance change in the channel in presence and absence of an analyte. The typical FET device is constructed by using the source and drain electrodes, a semiconducting layer, an insulator gate oxide, and a gate electrode (Figure 2.7). The drain current (I_{DS}) is controlled via electric field by applied voltage (V_{GS}) across the gate electrode and source. Therefore, the conductivity of the layer depends upon the potential used crossways the gate and source and the corresponding (I_{DS}) is given by (Lee et al. 2017):

$$I_{DS} = \frac{C_\mu W}{L}\left[(V_{GS} - V_{TH}) - \frac{1}{2}V_{DS}\right]V_{DS} \qquad (2.23)$$

where, C is the capacitance of the gate insulator per unit area, μ is the charge carrier mobility in the channel, W/L is the width-to-length ratio of the channel, V_{GS} and V_{DS} are the applied gate–source and drain–source voltage, and Vth is the threshold voltage, respectively.

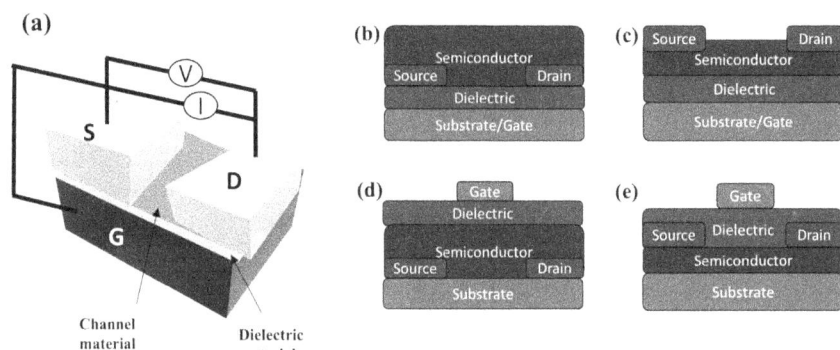

Figure 2.7 (a) The typical field effect transistor design, (b) Bottom gate/bottom contact, (c) Bottom gate/top contact, (d) Top gate/bottom contact, and (e) Top gate/top contact mode.

Generally, the linear region of I–V curve is obtained at lower drain–source voltages, which typically obeys the ohm's law. The FETs-based sensor performance is dependent mainly on the channel material and its properties like work function, charge-carrier movement, and band-gap. If there is a mismatch in the work function among the metal electrode and layer, it would degrade the carrier injection efficiency due to formation of Schottky barrier. Most common types of metals such as Cr/Au and Al/Au are preferred. For *p*-type material, an Au is used due to its larger work function whereas, for *n*-type material, the metals with lesser work function (Ni, Al, Cr) can be selected. To enhance the sensitivity, larger carrier mobility is considered. More importantly, the band gap of the sensing layer also decides the sensor performance. The sensitivity first increases and later falls with the decline in band gap (Meng et al. 2019; Yoshizumi 2017).

In recent years, numerous FETs are designed for chemical sensing such as catalytic-gate FETs, suspended-gate FETs (SGFETs), solid electrolyte-based FETs, and nanomaterial-based FETs (Pawar et al. 2021; Zhang et al. 2021). The FETs performance can be improved by utilizing the 2D materials. More studies need to be focused on simple device fabrication with high sensitivity, selectivity, stability and large-scale manufacturing.

2.7   QUARTZ CRYSTAL MICROBALANCE

The fundamental of quartz crystal microbalance can be well understood by Sauerbrey theory. The sensing mechanism of QCM is dependent upon the piezoelectric effect. In order to realize this effect, different cuts can be given to quartz crystal at different angles, and mostly typical types like AT-cut, BT-cut, FC-cut, SC-cut, etc., are used. Mostly, AT-cut is preferred due to suitable frequency and temperature features and typically used in thickness-shear mode. Also, zero-temperature-coefficient point lies around environmental temperature. Further, the thickness becomes lower with the increase in fundamental frequency of the crystal which leads to a lowering in quality factor of the crystal vibration. Consequently, the fundamental frequency of the QCM device normally lies within the 1–10 MHz (Dunham et al. 1995; O'Sullivan and Guilbault 1999). Typical frequency nature of QCM is shown in Figure 2.8.

The change in resonant frequency (ΔF) is related to primary frequency (F) of the crystal, mass of deposited material (M) in gram, and area (A) of electrode surface in cm^2. The change in frequency is given as follows (Dunham et al. 1995):

$$\Delta F = -2.3 \times 10^6 F^2 \frac{M}{A} \qquad (2.24)$$

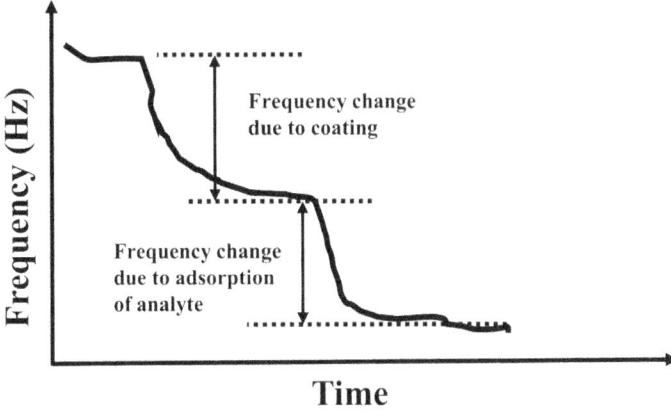

Figure 2.8 Typical quartz crystal microbalance-based sensor characteristics.

For QCM, an oscillating electric field through the device forms the acoustic wave which travels over the crystal and gets matched with the minimum impedance at a condition where the thickness must be a multiple of the half wavelength of the acoustic wave. The QCM works in shear mode, where acoustic wave propagation is perpendicular to the crystal surface. Therefore, the oscillating frequency is given by:

$$F = \frac{N}{a} \tag{2.25}$$

This equation can be modified further as:

$$\Delta F = -N\frac{\Delta a}{a^2} \tag{2.26}$$

From above equations:

$$\frac{\Delta F}{F} = -\frac{\Delta a}{a} \tag{2.27}$$

As the thickness of the crystal plate is related to the mass (M) of the crystal, area (A) of the crystal and density (ρ) of the crystal and is given by:

$$a = \frac{M}{A\rho} \tag{2.28}$$

$$\Delta a = \frac{\Delta M}{A\rho} \text{ for finite change}$$

$$\frac{\Delta F}{\Delta M} = -\frac{F}{M} \tag{2.29}$$

$$\Delta F = -k\Delta M$$

This is valid for a small changes in the mass only and for a large amount of change in mass thus, equation becomes invalid. More importantly, this equation shows that the sensitivity ($\Delta F/\Delta M$) of the sensor is directly proportional to the oscillating frequency of the crystal.

Now considering the shear modulus of the crystal, the above equation is modified as (Fauzi et al. 2021):

$$\Delta F = -\frac{2F^2}{\sqrt{\rho_q \mu_q}} \frac{\Delta M}{A} \tag{2.30}$$

where, μ_q and ρ_q are the shear modulus (g cm^{-1} s^{-2}) and density of quartz crystal (g cm^{-3}).

Now, when the crystal is coated with the thicker material, then the above equation needs to be corrected. When a crystal is immersed into a liquid-solution, the coupling between them changes the oscillation frequency of the crystal. The change in frequency is relative to the $(\rho\eta)^{1/2}$ where ρ_l and η_l are liquid density and viscosity and is given by (Dunham et al. 1995):

$$\Delta F = -\frac{F^{3/2} \rho_l (\eta_l)^{1/2}}{\sqrt{\pi \rho_q \mu_q}} \tag{2.31}$$

When the crystal is coated on both sides, the scale of ΔF becomes double. However, there might be a problem of liquid conductance shunting the crystal, and therefore, makes it more complicated. So, the QCM designs permit to coat one side only for better performance.

2.8   CONCLUSION AND OUTLOOK

In this chapter, we discussed the fundamental principles of various chemical detection methods. The theory and working principles of these techniques are discussed. The key highlights including influencing factors are covered. The optical fiber method is suitable due to its fast and miniaturized nature. The real time and distributed sensing can be possible by this technique. The chemiresistor approach is highly suitable due to simplicity, low-cost and reliability. The electrochemical method is widely used to extract the multiple information like analytical, physical and chemical parameters and the reaction kinetics. Capacitive method is also a very useful tool to measure the chemical species by measuring the capacitance. The FETSs are attracted due to their amplifying nature, low-cost fabrication and efficiency. The mechanism of quartz crystal microbalance technique in which the change in mass of the crystal in presence of chemical species is also discussed. These sensing methods offer great potential in the chemical sensing field due to their simple and easy fabrication, the assistance of advanced materials deposition techniques, and mass-production capability. These methods provide a real-time online monitoring of various chemical and biological species.

REFERENCES

Beć, Krzysztof Bernard, Justyna Grabska, and Christian Wolfgang Huck. 2020. Physical principles of infrared spectroscopy. pp. 1–43. *In*: Daniel Cozzolino (ed.). Infrared Spectroscopy for Environmental Monitoring. Comprehensive Analytical Chemistry, Vol. 98. Elsevier. doi:https://doi.org/10.1016/bs.coac. 2020.08.001.

Cao, PeiJiang, YongZhi Cai, Dnyandeo Pawar, S.T. Navale, Ch.N. Rao, Shun Han, WangYin Xu, et al. 2020. Down to ppb level NO_2 detection by ZnO/rGO heterojunction based chemiresistive sensors. Chemical Engineering Journal 401: 125491. doi:https://doi.org/10.1016/j.cej.2020.125491.

Cao, PeiJiang, XingGao Gui, Dnyandeo Pawar, Shun Han, WangYing Xu, Ming Fang, XinKe Liu, et al. 2021. Highly ordered mesoporous V_2O_5 nanospheres utilized chemiresistive sensors for selective detection of xylene. Materials Science and Engineering: B 265: 115031. doi:https://doi. org/10.1016/j.mseb.2020.115031.

Das, Sagnik, Subhajit Mojumder, Debdulal Saha and Mrinal Pal. 2022. Influence of major parameters on the sensing mechanism of semiconductor metal oxide based chemiresistive gas sensors: a review focused on personalized healthcare. Sensors and Actuators B: Chemical 352: 131066. doi:https://doi. org/10.1016/j.snb.2021.131066.

Devi, S., Raju K. Gupta, A.K. Paul, Vinay Kumar, Abhay Sachdev, P. Gopinath, and S. Tyagi. 2018. Ethylenediamine mediated luminescence enhancement of pollutant derivatized carbon quantum dots for intracellular trinitrotoluene detection: soot to shine. RSC Advances 8(57): 32684–32694. The Royal Society of Chemistry. doi:10.1039/C8RA06460A.

Dunham, Glen C., Nicholas H. Benson, Danuta Petelenz and Jiri. Janata. 1995. Dual quartz crystal microbalance. Analytical Chemistry 67(2): 267–272. American Chemical Society. doi:10.1021/ac00098a005.

Elosua, Cesar, Francisco Javier Arregui, Ignacio Del Villar, Carlos Ruiz-Zamarreño, Jesus M. Corres, Candido Bariain, Javier Goicoechea, et al. 2017. Micro and nanostructured materials for the development of optical fibre sensors. Sensors. 17(10): 2312. https://doi.org/10.3390/s17102312.

Fauzi, Fika, Aditya Rianjanu, Iman Santoso and Kuwat Triyana. 2021. Gas and humidity sensing with quartz crystal microbalance (QCM) coated with graphene-based materials—a mini review. Sensors and Actuators A: Physical 330: 112837. doi:https://doi.org/10.1016/j.sna.2021.112837.

Fleming, Karen G. 2017. Fluorescence theory. pp. 647–653. *In*: David Koppenaal, George E. Tranter and John C. Lindon (eds). Encyclopedia of Spectroscopy and Spectrometry, 3rd Ed. Oxford: Academic Press. doi:https://doi.org/10. 1016/B978-0-12-803224-4.00357-5.

Gehlen, Marcelo H. 2020. The centenary of the stern-volmer equation of fluorescence quenching: from the single line plot to the SV quenching map. Journal of Photochemistry and Photobiology C: Photochemistry Reviews 42: 100338. doi:https://doi.org/10.1016/j.jphotochemrev.2019.100338.

Gilbert, P.U.P.A. 2022. Physics in the Arts, 3rd Ed. (Chapter 2: Reflection and refraction, pp. 15–42). Academic Press. doi:https://doi.org/10.1016/B978-0-12-824347-3.00002-9.

Hasan, Anwarul, Nadir Mustafa Qadir Nanakali, Abbas Salihi, Behnam Rasti, Majid Sharifi, Farnoosh Attar, Hossein Derakhshankhah, et al. 2020. Nanozyme-based sensing platforms for detection of toxic mercury ions: an alternative approach to conventional methods. Talanta 215: 120939. doi:https://doi.org/10.1016/j.talanta.2020.120939.

Houck, Max M. and Siegel, Jay A. 2015. Fundamentals of Forensic Science, 3rd Ed. (Chapter 5: Light and matter, pp. 93–119). San Diego: Academic Press. doi:https://doi.org/10.1016/B978-0-12-800037-3.00005-4.

House, J.E. 2018. Fundamentals of Quantum Mechanics, 3rd Ed. (Chapter 11: Molecular spectroscopy, pp. 271–296). Academic Press. doi:https://doi.org/10.1016/B978-0-12-809242-2.00011-5.

Hoyos-Arbeláez, Jorge, Mario Vázquez and José Contreras-Calderón. 2017. Electrochemical methods as a tool for determining the antioxidant capacity of food and beverages: a review. Food Chemistry 221: 1371–1381. doi:https://doi.org/10.1016/j.foodchem.2016.11.017.

Jespersen, N. 2006. General principles of spectroscopy and spectroscopic analysis. pp. 111–155. In: S. Ahuja and N. Jespersen (eds). Modern Instrumental Analysis, Vol. 47. Series: Comprehensive Analytical Chemistry. Elsevier. doi:https://doi.org/10.1016/S0166-526X(06)47005-7.

Koo, Won-Tae, Jang Ji-Soo and Kim, Il-Doo. 2019. Metal-Organic frameworks for chemiresistive sensors. Chem 5(8): 1938–1963. doi:https://doi.org/10.1016/j.chempr.2019.04.013.

Korotcenkov, G. 2007. Metal oxides for solid-state gas sensors: what determines our choice? Materials Science and Engineering: B 139(1): 1–23. doi:https://doi.org/10.1016/j.mseb.2007.01.044.

Kummer, Adrian M., Andreas Hierlemann and Henry Baltes. 2004. Tuning sensitivity and selectivity of complementary metal oxide semiconductor-based capacitive chemical microsensors. Analytical Chemistry 76(9): 2470–2477. American Chemical Society. doi:10.1021/ac0352272.

Lee, Yoon Ho, Moonjeong Jang, Moo Yeol Lee, O Young Kweon and Joon Hak Oh. 2017. Flexible field-effect transistor-type sensors based on conjugated molecules. Chem 3(5): 724–63. doi:https://doi.org/10.1016/j.chempr. 2017.10.005.

Lv, Hualiang, Zhihong Yang, Hongge Pan and Renbing Wu. 2022. Electromagnetic absorption materials: current progress and new frontiers. Progress in Materials Science 127: 100946. doi:https://doi.org/10.1016/j.pmatsci.2022.100946.

Majhi, Sanjit Manohar, Ali Mirzaei, Hyoun Woo Kim, Sang Sub Kim and Tae Whan Kim. 2021. Recent advances in energy-saving chemiresistive gas sensors: a review. Nano Energy 79: 105369. doi:https://doi.org/10.1016/j.nanoen.2020.105369.

Meng, Zheng, Robert M. Stolz, Lukasz Mendecki and Katherine A. Mirica. 2019. Electrically-transduced chemical sensors based on two-dimensional nanomaterials. Chemical Reviews 119(1): 478–598. American Chemical Society. doi:10.1021/acs.chemrev.8b00311.

Militký, J. 2011. Fundamentals of soft models in textiles. pp. 45–102. *In*: A. Majumdar (ed.). Soft Computing in Textile Engineering. Woodhead Publishing. doi:https://doi.org/10.1533/9780857090812.1.45.

O'Sullivan, C.K. and G.G. Guilbault, 1999. Commercial quartz crystal microbalances—theory and applications. Biosensors and Bioelectronics 14(8): 663–670. doi:https://doi.org/10.1016/S0956-5663(99)00040-8.

Panasyuk, Tatiana L., Vladimir M. Mirsky, Sergey A. Piletsky and Otto S. Wolfbeis. 1999. Electropolymerized molecularly imprinted polymers as receptor layers in capacitive chemical sensors. Analytical Chemistry 71(20): 4609–4613. American Chemical Society. doi:10.1021/ac9903196.

Patel, Bhavik A. 2020. Electrochemistry for Bioanalysis (Chapter 2: Amperometry and potential step techniques, pp. 9–26). Elsevier. doi:https://doi.org/10.1016/B978-0-12-821203-5.00009-9.

Pawar, Dnyandeo and S.N. Kale 2016. Birefringence manipulation in tapered polarization-maintaining photonic crystal fiber mach-zehnder interferometer for refractive index sensing. Sensors and Actuators A: Physical 252: 180–184. doi:https://doi.org/10.1016/j.sna.2016.10.032.

Pawar, Dnyandeo, Ch.N. Rao, Ravi Kant Choubey and S.N. Kale. 2016. Mach-zehnder interferometric photonic crystal fiber for low acoustic frequency detections. Applied Physics Letters 108(4): 41912. American Institute of Physics. doi:10.1063/1.4940983.

Pawar, Dnyandeo, Rohini Kitture and S.N. Kale. 2017. ZnO coated fabry-perot interferometric optical fiber for detection of gasoline blend vapors: refractive index and fringe visibility manipulation studies. Optics & Laser Technology 89: 46–53. doi:https://doi.org/10.1016/j.optlastec.2016.09.038.

Pawar, Dnyandeo, B.V. Bhaskara Rao and S.N. Kale. 2018. Fe_3O_4-Decorated graphene assembled porous carbon nanocomposite for ammonia sensing: study using an optical fiber fabry–perot interferometer. Analyst 143: 1890–1898. Royal Society of Chemistry. doi:10.1039/c7an01891f.

Pawar, Dnyandeo and N. Kale. 2019. A review on nanomaterial-modified optical fiber sensors for gases, vapors and ions. Microchimica Acta 186(4): 253. doi:10.1007/s00604-019-3351-7.

Pawar, Dnyandeo, Shankar Gaware, Ch.N. Rao, Rajesh Kanawade and Peijiang Cao. 2021. Gas sensors-based on field-effect transistors. pp. 355–375. *In*: Inamuddin, R. Boddula, Abdullah M. Asiri and Md. M. Rahman (eds). Green Sustainable Process for Chemical and Environmental Engineering and Science: Solid State Synthetic Methods. Elsevier. doi:https://doi.org/10.1016/B978-0-12-819720-2.00020-5.

Shibuya, T. and K. Kawase. 2013. Terahertz applications in tomographic imaging and material spectroscopy: a review. pp. 493–509. *In*: S. Daryoosh (ed.). Handbook of Terahertz Technology for Imaging, Sensing and Communications: A volume in Woodhead Publishing Series in Electronic and Optical Materials. Woodhead Publishing. doi:https://doi.org/10.1533/9780857096494.3.493.

Tsouti, V., C. Boutopoulos, I. Zergioti and S. Chatzandroulis. 2011. Capacitive microsystems for biological sensing. Biosensors and Bioelectronics 27(1): 1–11. doi:https://doi.org/10.1016/j.bios.2011.05.047.

Ugale, Ashok D., Resham V. Jagtap, Dnyandeo Pawar, Suwarna Datar, S.N. Kale and Prashant S. Alegaonkar. 2016. Nano-carbon: preparation, assessment, and applications for NH_3 gas sensor and electromagnetic interference shielding. RSC Advances 6(99): 97266–97275. The Royal Society of Chemistry. doi:10.1039/C6RA17422A.

Urrutia, Aitor, Javier Goicoechea and Francisco J. Arregui. 2015. Optical fiber sensors based on nanoparticle-embedded coatings. Journal of Sensors 2015: 805053. doi:10.1155/2015/805053.

Wang, Xu-dong and Wolfbeis Otto S. 2016. Fiber-optic chemical sensors and biosensors (2013–2015). Analytical Chemistry 88(1): 203–227. American Chemical Society. doi:10.1021/acs.analchem.5b04298.

Wypych, George. 2015. Mechanisms of UV stabilization. pp. 37–65. *In*: George Wypych (ed.). Handbook of U.V. Degradation and Stabilization, 2nd Ed. ChemTec Publishing, Toronto, Ontario. doi:https://doi.org/10.1016/B978-1-895198-86-7.50005-X.

Yao, Ming-Shui, Wen-Hua Li and Gang Xu. 2021. Metal–organic frameworks and their derivatives for electrically-transduced gas-sensors. Coordination Chemistry Reviews 426: 213479. doi:https://doi.org/10.1016/j.ccr.2020.213479.

Yoshizumi, Toshihiro and Yuji Miyahara. 2017. Field-effect transistors for gas sensing. pp. 1–192. *In*: Momčilo Pejović and Milić M. Pejovic (eds). Different Types of Field-Effect Transistors, Theory and Applications. IntechOpen, London. doi:10.5772/intechopen.68481.

Zhang, John X.J. and K. Hoshino. 2019. Molecular Sensors and Nanodevices: Principles, Designs and Applications in Biomedical Engineering (Chapter 4: Electrical transducers: electrochemical sensors and semiconductor molecular sensors, pp. 181–230). Academic Press. doi:https://doi.org/10.1016/B978-0-12-814862-4.00004-1.

Zhang, Pan, Yin Xiao, Jingjing Zhang, Bingjie Liu, Xiaofei Ma and Yong Wang. 2021. Highly sensitive gas sensing platforms based on field effect transistor—a review. Analytica Chimica Acta 1172: 338575. doi:https://doi.org/10.1016/j.aca.2021.338575.

Chapter **3**

Materials for Chemical Detection

3.1 INTRODUCTION

Since many years, materials are playing a significant role in chemical sensing. Numerous materials have been investigated for the detection of various chemical parameters (like gas, ion, pH, humidity, and bio-chemical species). At present, nanotechnology has been growing tremendously, and has therefore, attracted considerable interest in the development of nanostructured materials with extraordinary electronic, chemical, and optical properties for chemical detection. The ability to tune surface to volume ratio, incorporation of surface defects, and activation energy toward the analyte are the key features of a chemical sensor. In chemical sensing, the selection of proper materials with controllable size, shape, morphology, surface area, and oxygen vacancies is very crucial (Das et al. 2022; Rao et al. 2002; Yoon et al. 2021). Also, the materials should get the least interference by the surrounding parameters (mostly temperature, humidity, and interfering agents), which might affect the sensor performance (Das et al. 2022). Overall, material's physical and chemical properties play a significant role in deciding the chemical sensor performance. Therefore, the choice of an appropriate material is very important which could fulfill the demands of the sensor.

Till now various nanostructured materials in the form of 0D, 1D, 2D and 3D have been explored for chemical sensing. The most promising materials like metal oxides (MOXs), carbon nanotubes, 2D materials such as graphene oxide, transition metal chalcogenides, black phosphorus, MXenes, etc., hybrid composites such as MOX-metals,

MOX-2D, polymeric materials, and MOX-polymers have been extensively investigated for the detection of various chemical parameters and have displayed great sensing performance (Deshmukh et al. 2020; Kausar 2022; Lange et al. 2008; Meng et al. 2019; Stock and Biswas 2012; Wilson and Yoffe 1969; Wackerlig and Lieberzeit 2015; Yoon et al. 2021). This chapter summarizes the details of recent nano-materials and their key features along with their limitations.

3.2 NANOSTRUCTURED MATERIALS AND THEIR DERIVATES

3.2.1 Metal Oxides

Metal oxides (MOX) are produced due to the unique combination of oxide ions of metal and oxygen. The metal oxides are widely studied and very favorable in the chemical sensing domain due to their extraordinary electrical properties, simple and low-cost synthesis, high repeatability, selective, and good long-life constancy. MOXs easily design into any size, shape, surface area, and flexibility and are hence widely adopted for chemical sensing. The conductivity of metal oxides is dependent upon the crystallite size, morphological structure, dopant, and operation temperature, etc. (Comini and Zappa 2020; Comini 2013; Agnihotri et al. 2021; Mondal and Gogoi 2022). Therefore, the MOXs exhibit varied electrical properties ranging from best insulators to narrow-band gap semiconductors to metals to finally superconductors. The following Table 3.1 depicts the type of MOXs with their type of conductivity. So, these are divided into two types as follows (Korotcenkov 2007):

1. Transition-metal oxides
2. Non-transition-metal oxides
   (a) Pre-transition-metal oxides
   (b) Post-transition-metal oxides

Generally, MOXs sensing mechanism depends upon the variation in resistance of the material when it comes in contact with any chemical species. For example, in the case of n-type material, the depletion width decreases upon interaction with reducing gases and vice versa. Whereas for p-type material, the depletion width increases upon interaction with oxidizing gases and vice versa (Korotcenkov and Cho 2017; Mondal and Gogoi 2022). Table 3.1 shows the nature and properties of the materials depending upon the type of the species. However, most of the MOXs are well operated at an elevated temperature because at high temperature, it promotes conductivity and surface reactions. By

Table 3.1 Type of conductivity of some sensing materials along with the change in their conductivity in presence of reducing and oxidizing gas.

Material	Type of conductivity		
	n	**p**	**n, p**
Metal oxides	ZnO, In$_2$O$_3$, SnO$_2$, MgO, TiO$_2$, ZrO$_2$, CaO, V$_2$O$_5$, Nb$_2$O$_5$, Ta$_2$O$_5$, MoO$_3$, WO$_3$, Al$_2$O$_3$, Ga$_2$O$_3$, SrTiO$_3$, SrTiFeO$_3$	Co$_3$O$_4$, NiO, Bi$_2$O$_3$, Y$_2$O$_3$, La$_2$O$_3$, CeO$_2$, Mn$_2$O$_3$, PdO, Ag$_2$O, Sb$_2$O$_3$, TeO$_2$	CuO, Fe$_2$O$_3$, HfO$_2$, Cr$_2$O$_3$
Semiconductors	GaN, SiC, diamond		Si, InP, GaAs
Material property	**n**	**p**	**Type of gas**
Conductivity	Increases	Decreases	Reducing gas
Carrier concentration	Increases	Decreases	
Conductivity	Decreases	Increases	Oxidizing gas
Carrier concentration	Decreases	Increases	

employing various strategies like mixing with polymer and metals, the working temperature can be decreased.

The semiconductor metal oxides can be clubbed together with different materials (shown in Figure 3.1) like carbon, metals, and polymers to form a unique combination for the detection of any chemical species. The metal oxides in various forms such as spheres, flowers, rods, particles, dendrite shapes have been synthesized and successfully employed for chemical sensing (Li et al. 2019; Xie et al. 2022).

3.2.1.1 Various types of MOX-based sensors

Based on the combination of the various metal oxides, different structures (heterojunction, homojunction, Schottky contact) are realized. When two materials with different fermi energies come in contact with each other, it leads to formation of an interfacial region. Upon interaction of chemical species with the material, it changes accordingly and thus, variations in the electrical resistance of the composite, and thereby decide the performance of the sensor. The different types of nanocomposites can be realized as follows.

1. Metal oxides modified with metal oxides
   (p-n heterojunction: CuO/ZnO, NiO/SnO$_2$, etc.)
   (n-n heterojunction: SnO$_2$@TiO$_2$, ZnO/In$_2$O$_3$, etc.)
   (p-p heterojunction: CuO and NiO, Co$_3$O$_4$/CuO, etc.)

2. Metal oxides modified with metal nanoparticles
   (WO$_3$/Pt, Pd/ZnO, etc.)

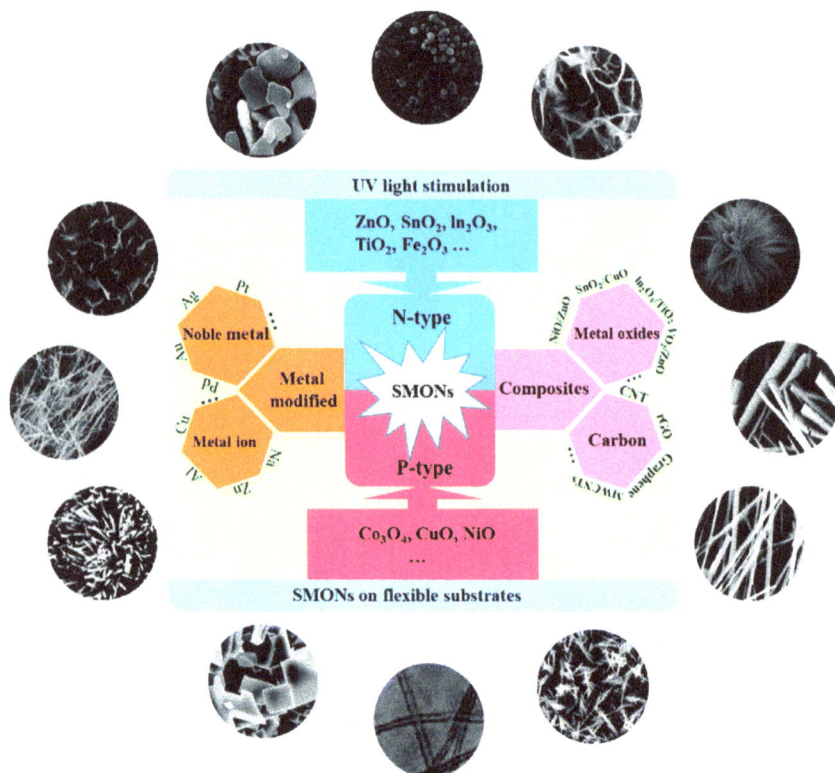

Figure 3.1 The graphic represents numerous semiconducting MOXs and their synthesis approach. [Reproduced with permission (Li et al. 2019) Copyright 2019, Royal Society of Chemistry].

3. Metal oxides modified graphene-based materials
    (ZnO/rGO, TiO_2/rGO, etc.)

4. Metal oxides modified 2D TMDs materials
    (MoO_3/MoS_2, SnO_2-SnS_2, etc.)

5. Metal oxides modified MOF
    ($ZnO-Co_3O_4/Zn$-ZIF-67, $Co_3O_4/In_2O_3/Co$—MIL—68(In), etc.)

6. Metal oxides with polymer
    (SnO_2/PPy, $PANI/Fe_2O_3$, etc.)

The literature study indicates that the MOXs based materials need to address certain challenges (Chowdhury and Bhowmik 2021; Korotcenkov and Cho 2017; Li et al. 2019) such as (1) controlling the thickness and lateral size during the top-down and bottom-up approaches, (2) specific attention needs to be paid to the kinetics and direction of the materials confined growth, and (3) understanding and controlling the surface defects and oxygen vacancies in order to fabricate an ideal chemical sensor.

3.2.2 2D Materials

2D materials are having thicknesses in the range of nanometer and adjacent sizes in the order of around centimeters. This is a class of materials widely investigated for chemical detection because of their fascinating electro-optic and morphological properties. Material like graphene and its analogues materials namely transition metal chalcogenides, hexagonal shape boron nitride, black phosphorus, MXene and MOFs are belonging to this group. These materials exhibit mesmerizing properties like high surface area, atom level thickness, high electrical conductivity, high active sites on the surface, ability to tune electrical properties and binding nature (Meng et al. 2019). From a device point of view, these materials exhibit low-power consumption, high mobile charge carriers, low-weight, high stretchability and flexibility, good mechanical stability, excellent optical transparency, and good environmental stability, etc. The discussion on each material is provided in the following section.

3.2.3 Graphene and its Derivatives

Graphene is very attractive for chemical sensing due to its one atom thick layer which increases its surface area multi-fold times, and therefore, has a great ability to detect a single molecule. The graphene exhibits a massive charge-carrier motion around 2×10^5 cm^2 V^{-1} s^{-1} and a large charge-carrier density around 10^{12} cm^{-2} at RT and thus, exhibits a low resistivity around 10^{-6} Ω. Other properties include an enormous specific surface area of around 2630 m^2 g^{-1} with the Young's modulus of around 1 TPa (Meng et al. 2019; Yang et al. 2016). As the layer of graphene increases, it tends to form a graphite structure. The graphene can be synthesized by using various methods like exfoliation of graphite, chemical vapor deposition (CVD), organic preparation, Scotch tape and micromechanical cleaving, etc.

The graphene derivatives such as graphene oxide can be easily synthesized by a solution-phase exfoliation having oxidation of graphite in presence of a strongly acidic medium. The synthesis of graphene oxide is shown in Figure 3.2. The attractive features of graphene oxide are the existence of abundance of functional groups on its surface. However, due to sp^3 C–C bonds, it displays an insulating property. The electrical property of these materials can be regained by reducing the GO either thermally, electrically or chemically and thus, reduced graphene oxide can be formed (Ajala et al. 2022).

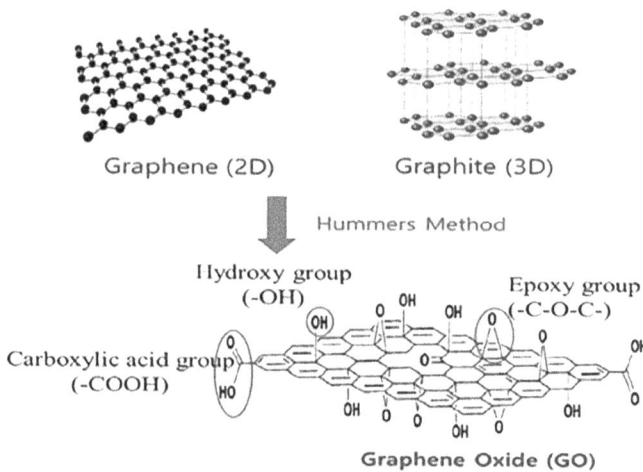

Graphene (2D) Graphite (3D)

Hummers Method

Hydroxy group (-OH)

Epoxy group (-C-O-C-)

Carboxylic acid group (-COOH)

Graphene Oxide (GO)

Figure 3.2 The synthesis of GO from graphite by Hummer's method. [Reproduced with permission (Ajala et al. 2022) Copyright 2022, Elsevier].

The graphene and rGO have been extensively used in chemical sensing. The attractive features (Ajala et al. 2022; Kausar 2022) of these materials include such as (1) very high specific surface area, and therefore, every carbon atom is contributing to the interaction, (2) ability to interact by van-der Waals force, electron transferal, or covalent attachment, (3) easily forming a heterostructure with other materials like MOX, metals, and 2D materials, etc. (4) the electrical signal shows high SNR ratio due to fewer surface defects, (5) can be generated into flexible structures with high stretchability and bending property.

However, these materials also exhibit certain limitations (Aiswaria et al. 2022; Ghosh et al. 2022) such as (1) The intrinsic graphene may exhibits poor sensitivity due to its inert chemical nature and weak Vander-walls interaction with analyte. To realize a good performance of the sensor, the surface can be modified by dopants or making composites by using metallic, MOXs and polymeric materials. (2) Regarding GO, a detailed study of its reduction, chemical alteration, controlling oxygenated functionality groups and development of composites is necessary. (3) The selectivity needs to improve by the addition of other materials. (4) Expensive fabrication set up to obtain a good quality of graphene films, and (5) Need to understand detailed surface interaction and lattice vacancies of these materials.

3.2.4 Transition Metal Dichalcogenides

Transition metal dichalcogenides (TMDCs) belong to the inorganic materials category by a general formula as X-M-X, where 'M' is a hexavalent transition metal ion (W, Ta, Zr, etc.), whereas X is known as divalent chalcogen transition metal atoms (S, Se, and Te) (shown in Figure 3.3). Generally, these materials are observed in the form of layered structures with strong bonds existing inside layers and considerably weaker in the nearby layers. In 1986, a single-layer of MoS_2 was reported by the lithium intercalation (Joensen et al. 1986).

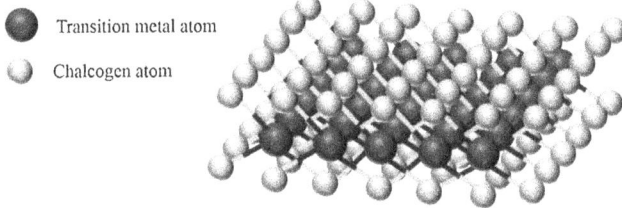

Figure 3.3 An indication of the transition metal dichalcogenide materials in the periodic table and their corresponding layered crystallite structure. [Reproduced with permission (Kumar et al. 2020) Copyright 2020, Elsevier].

The electrical properties of TMDCs considerably vary from insulating HfS_2, semiconducting MoS_2 and WS_2, semi-metallic TeS_2, and metallic NbS_2 and TaS_2 (Wilson and Yoffe 1969). An exfoliation of bulk TMDCs significantly changes $s–p_z$ orbital bonding between nearby layers which further widens the band-gap. The external layer of TMDCs contains chalcogen atoms with a lone pair and so, it gives a well compounded constancy. TMDCs materials own extraordinary properties like direct band-gap, low level thickness, and good spin-orbit coupling. Another feature includes easy functionalization, very high surface area, and the ability to tune bandgap and nanosheet dimensionality (Anju and Mohanan 2021; Kumar et al. 2020; Ping et al. 2017). MoS_2 and WS_2 materials only exist in the layer form whereas the bulk TMDCs are synthesized by bottom-up approach. The TMDCs including monolayer and multi-layer are synthesized via top-down technique (Anju and Mohanan 2021). The list of top-down and bottom-up synthesis methods details are as follows:

The top-down method includes:

1. Micromechanical exfoliation
2. Liquid exfoliation
3. Hybrid exfoliation
4. Etching

The bottom-up method includes

1. CVD
2. Wet chemical synthesis

However, the following are the certain challenges that must be overcome in order to realize a superior material (Kumar et al. 2020; Meng et al. 2019). The even development of monolayer TMDCs is always difficult. It was observed that most of the TMDCs compounds exhibited non-uniform layer of distribution when using the liquid exfoliation methods. In the case of ion-intercalation technique, it is difficult to balance the layer dissipation and stability. Therefore, there is difficulty in obtaining the specific layer multiplicity and tailoring edge functionality. It is a relatively new class of materials, and therefore, physico-chemical, dispersion property, controlling uniformity of size and shape dispersal, toxicity and environmental stability need to be studied further.

3.2.5   MXenes

At present, MXenes are a very new kind of materials in 2D world reported in 2011 (Naguib et al. 2012). The MXenes (around 30 compounds reported) are represented by a formula as $M_{n+1}X_nT_x$. where, M signifies transition metallic materials like Mo, Hf, Ti, Ta, W, Cr, Sc, V, Nb, and Y; n: an integral number ranging from 1 and 3, X stands for carbon/nitrogen. T denotes surface termination groups like hydroxyl (OH), fluorine (F), chlorine (Cl), and oxygen (O), and x denotes the count of surface functional groups (Verger et al. 2019). In 2D material, transition metal nitrides, carbides, and carbonitrides are together mentioned as MXenes and widely used in various fields owing to their unique electric, optical, mechanical and biological properties.

MXenes possess various interesting physical, chemical, electrical, and optical properties. The oxygen and extra OH MXene terminations disturb its surface hydrophilic property. MXenes are mechanically stable because of strong covalent bonding among transition metallic and carbon/nitrogen. The Young's modulus of MXenes is around 17.3 and 333 GPa, which is much higher among the 2D materials category (Kailasa et al. 2021). MXene hydrogel (Ti_3C_2/PAM NC) is shown high elongation and bending almost by more than 10 times and 180°, respectively (Kailasa et al. 2021). MXenes show remarkable optical properties, and therefore, are widely used in photocatalysis, photoconductive and opto-electric sensors and have also been tested to develop a highly conducive transparent electrode. $Ti_3C_2T_x$ MXene film transfers around 91.2% light in UV-vis wavelength (Deshmukh et al. 2020). If these are synthesized without terminations, it could act as an electrical conductor, and by functionalizing their surfaces, the metallic property can be easily transformed into semiconducting nature. For example, $Ti_3C_2T_z$ showed a large conductivity (9880 S cm^{-1}) (Khazaei et al. 2017). It also displays magnetic properties due to the occurrence of transition metals, functional surface and single-layer defects. If it is etched via CS and HF acidic solvents, the MXenes can be converted to paramagnetic and antiferromagnetic material. The grouping of carbon and nitrogen also plays a crucial part to decide the magnetic property and thus, (V_2C and V_2N) MXenes shows antiferromagnet and non-magnetic property (Deshmukh et al. 2020).

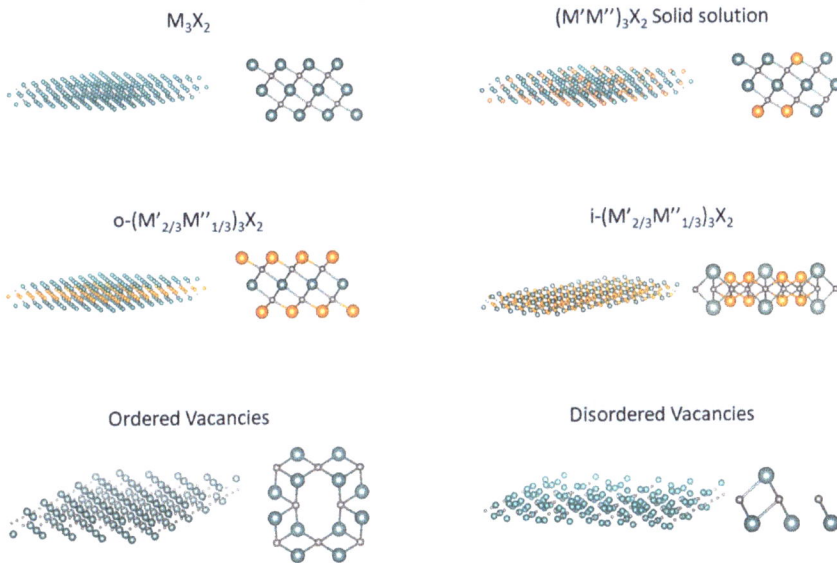

Figure 3.4   Various types of MXenes-based structures. [Reproduced with permission (Kailasa et al. 2021) Copyright 2021, Elsevier].

MXenes can be synthesized by using the following methods as shown in Figure 3.4 (Ashton et al. 2019; Kailasa et al. 2021):

The widely used top-down approach containing:

1. Etching method
2. Exfoliation process

The extensively used bottom-up approach includes:

1. CVD
2. Template method
3. Plasma enhanced pulsed laser deposition

MXenes and their nanocomposites are very promising in various fields. However, there is a scope to improve the material property for the fabrication of an ideal chemical sensor (Kailasa et al. 2021). The limitations can be addressed are (1) single-step reaction to synthesis of MXenes, (2) combined synthesis of many metals or nonmetals without etching solution, (3) tuning of porosity and functionalization of MXenes.

3.2.6   Metal Organic Frameworks

Metal-organic frameworks (MOFs) designed from metallic ions joined by di- or multi-topic organic linkers were first reported in 1990s (Yaghi et al. 1995). MOFs can be designed into 1D, 2D, 3D structures depending upon the class of the metallic and linkers utilized. Till now, around 20,000 different MOFs are reported and these materials are famous for their superior porosity and crystalline nature. These materials exhibit properties like ultra-high specific surface area, tunability in pore size distribution, highly active, and chemically and thermally stable, etc. (Jiao et al. 2019). However, they won't show very high electrical conductivity because of fewer charge carriers present and a large energy band gap. The electrical conductivity can be enhanced by incorporating metal oxides and polymers. By proper selecting (metallic and organic ligands), band gap of MOFs can be tuned systematically (Meng et al. 2019; Shi et al. 2021).

MOFs can be synthesized by the following methods. The synthesis methods and parameters are shown in Figure 3.5. Generally, the MOFs can be synthesized by commonly used efficient methods as follows (Stock and Biswas 2012).

1. Microwave-assisted thermal deposition
2. Interface assisted synthesis
3. Liquid phase epitaxy
4. Spray coating
5. Electrochemical synthesis
6. In-situ Synthesis
7. Vapor phase synthesis
8. Sonochemical Synthesis

The MOF stability is dependent upon many other parameters such as metallic ions, size, shape of organic ligands, metals and ligands and pore-size distribution, etc. Due to high porosity, the MOFs (low-valent metal ions) generally show poor mechanical strength. However, Zr-MOFs show better mechanical stability. The thermal stability of MOFs increases with high-valency species like Al^{3+}, Zr^{4+}, Ti^{4+} and changing the functionality. MOF films (HKUST-1 SURMOF) exhibit a smooth surface and good light transmittance, electrochromic properties (NBU-3 MOF), and photoluminescent (Ln^{3+}@MIL-100(In)), etc. (Garg et al. 2021; Shi et al. 2021).

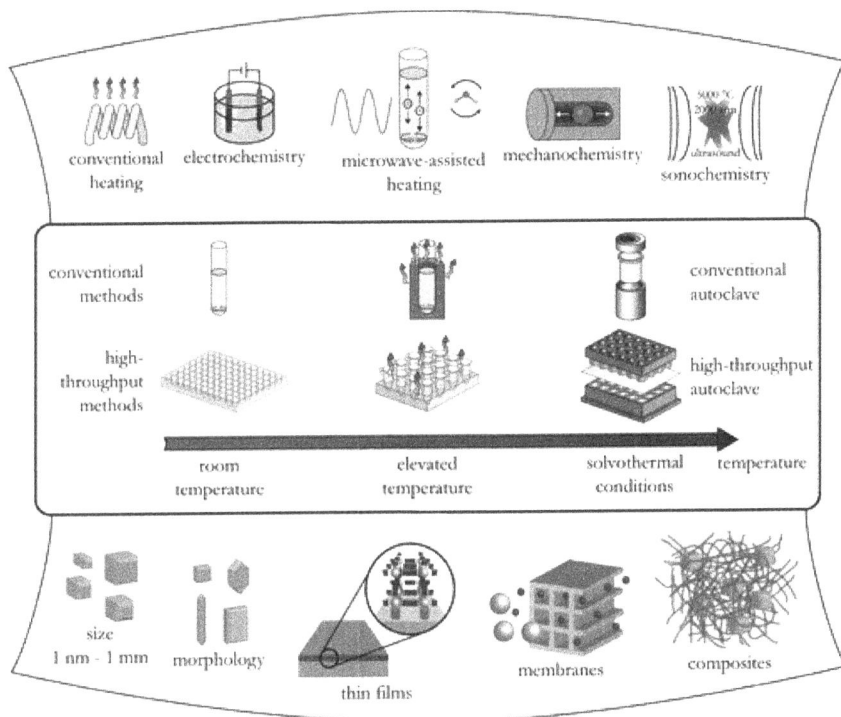

Figure 3.5 Overview of synthesis methods, possible reaction temperatures, and final reaction products in MOF synthesis. [Reproduced with permission (Stock and Biswas 2012) Copyright 2012, American Chemical Society].

The MOFs exhibit many interesting properties and have tremendous potential in industrial and environmental applications. However, there is a need to understand the surface chemistry of MOFs. Also, the synthesis of MOFs via a simple, and low-cost technique is required. There might be difficulty in obtaining the uniform MOF film on the substrate due to their poor heterogeneous nucleation on the substate. The LPE method provides high crystallinity and uniformity and thus, is widely adopted for good quality MOF films but it is time-consuming. Developing a cost-effective, simple method for the preparation of homogeneous and polycrystalline films on the substrate is a challenge. Also, the growth-time of MOF is quite lengthy and so, controlling parameters is necessary (Jiao et al. 2019; Meng et al. 2019; Shi et al. 2021).

3.2.7 Polymers

At present, polymers are the backbone of chemical sensing due to their fascinating and mesmerizing properties. The polymers have existed in various forms like conducting polymers, hydrogels, and molecular imprinted polymers (MIP), etc. These polymers exhibit high flexibility, biocompatibility, and long-term stability. Polymer-based hydrogels are hydrophilic crosslinked polymer networks that absorb and retain large amounts of water. It is highly sensitive to solvent composition, pH, temperature, electric field, and light. The structural properties of hydrogels can be enhanced by incorporating metal oxides, metals and carbon nanomaterials (Kailasa et al. 2022).

Conducting or conjugated polymers are poly-unsaturated compounds with all atoms being sp or sp^2 hybridized (as shown in Figure 3.6). Generally, it is insulating in nature (in uncharged condition) and depending upon the type of charge carriers like oxidizing (p-doping) or reducing (n-doping), their intrinsic conductivity arises. The dopant ions are generally incorporated into the polymer during the chemical or electrochemical polymerization process (Alvarez et al. 2011; Lange et al. 2008). The polymers have many interesting properties such as conducting polymers' own multiple merits like work at low temperature, simple and easy chemical alteration, and is being highly flexible, and therefore, widely utilized for chemical sensing. The most commonly used polymers are displayed in Figure 3.6.

The conducting polymers synthesis was first reported in 1970. The CPs can be synthesized by following methods (Alvarez et al. 2011).

1. Electrochemical polymerization
2. Chemical polymerization
3. Functionalized conjugated polymers

Certain conducting polymer are electrochemically stable (like neutral polyacetylene) and have less solubility as compared to saturated organic polymers. The optical and electronic properties are dependent upon the Conjugation length. Change in CP electronic band structure can lead to the difference in the electronic and optical properties of the polymers and thus, exhibit electrochromic and electroluminescence properties (Alvarez et al. 2011; Carvalho et al. 2018).

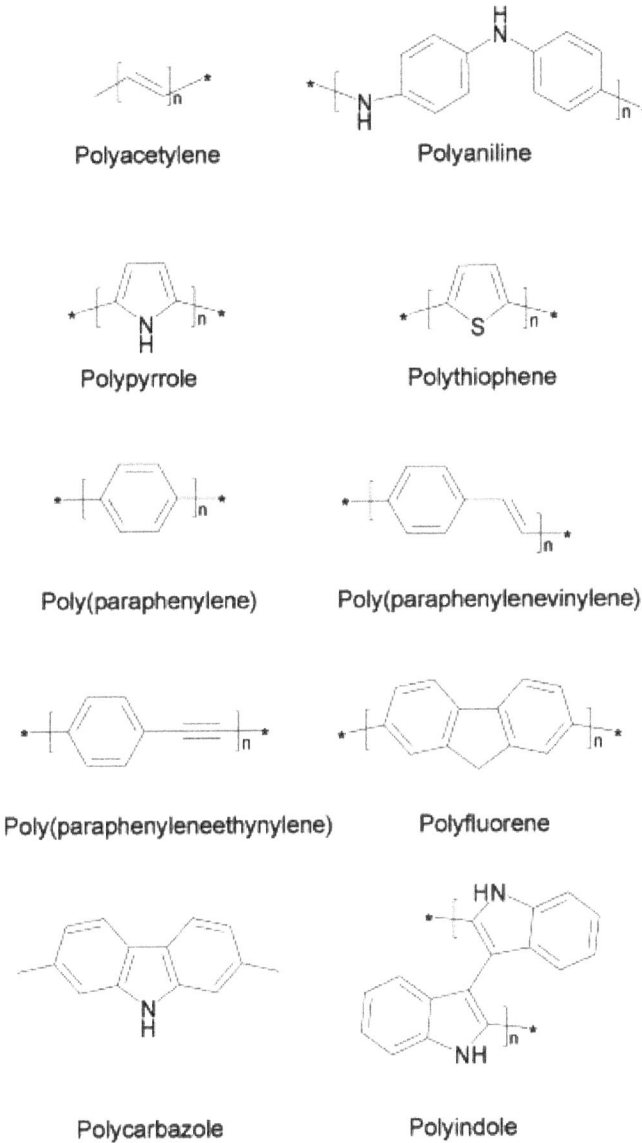

Polyacetylene   Polyaniline

Polypyrrole   Polythiophene

Poly(paraphenylene)   Poly(paraphenylenevinylene)

Poly(paraphenyleneethynylene)   Polyfluorene

Polycarbazole   Polyindole

Figure 3.6   Various types of conducting polymers. [Reproduced with permission (Lange et al. 2008) Copyright 2008, Elsevier].

3.3   CONCLUSION AND OUTLOOK

The sensing material is a vital part of a chemical sensor. This chapter reviews the most widely utilized materials such as semiconducting

metal oxides, 2D materials, and polymers, which are very effective for chemical sensing due to their superb physico-chemical and electro-optic properties. With growing nanotechnology, different types of morphologies can be realized with great control over their different properties. The metal oxides are very famous for chemical sensing due to their easy synthesis, low-cost, tuning ability, great compatibility with electronic interfaces, and easy to make nanocomposites. At present, several 2D nanomaterials are proving to be very effective owing to their huge surface area along with their excellent electrical properties and thus, in the future, these materials will be able to prove their value in scientific research and practical applications. The polymers are extensively synthesized and used for chemical sensing. These materials are low-cost, easy to synthesize, highly flexible, and can tune with other nanomaterials. The limitations of these materials along with the futuristic material designs are discussed.

The materials with a high ability to tune their physical, chemical, optical, and electrical properties are highly preferred in chemical sensing. Most of the material fabrication involves a multistep synthesis fabrication process and thus, it becomes time-consuming. The synthesis of well-ordered core–shell structures, synthesis of highly porous materials, and good repellent materials to humidity are challenging. The sensing material should be manufactured in an easy way and under control over its reproducibility in terms of its size, shape, powder, films, and composite forms, etc. While synthesizing the material, its homogeneity must be controlled for effective sensing performance. Also, compatibility with the standard electronic interface must be studied. The difficulty in controlling their toxicity and this varying from material to material adds to the complexity in sensor fabrication. Obtaining the multifunctional material is a great challenge as it involves great control over the surface chemistry of the materials. Special attention to tuning materials' properties for enhancing sensitivity and selectivity is also crucial. The viability of materials to sensor fabrications is very critical. The design of sensing material with high durability, reproducibility, and reusability must be considered.

REFERENCES

Agnihotri, Ananya S., Anitha Varghese and Nidhin, M. 2021. Transition metal oxides in electrochemical and bio sensing: a state-of-art review. Applied Surface Science Advances 4: 100072. doi:https://doi.org/10.1016/j.apsadv.2021.100072.

Aiswaria, P., Samsudeen Naina Mohamed, Singaravelu D. Lenin, Kathirvel Brindhadevi and Arivalagan Pugazhendhi. 2022. A review on graphene/ graphene oxide supported electrodes for microbial fuel cell applications: challenges and prospects. Chemosphere 296: 133983. doi:https:// doi.org/ 10.1016/j.chemosphere.2022.133983.

Ajala, O.J., J.O. Tijani, M.T. Bankole and A.S. Abdulkareem. 2022. A critical review on graphene oxide nanostructured material: properties, synthesis, characterization and application in water and wastewater treatment. Environmental Nanotechnology, Monitoring and Management 18: 100673. doi:https://doi.org/10.1016/j.enmm.2022.100673.

Alvarez, Adrian, Alfonso Salinas-castillo, Rosario Pereiro and Alfredo Sanz-medel. 2011. Fluorescent conjugated polymers for chemical and biochemical sensing 30(9): 1513–1525. doi:10.1016/j.trac.2011.04.017.

Anju, S. and P.V. Mohanan. 2021. Biomedical applications of transition metal dichalcogenides (TMDCs). Synthetic Metals 271: 116610. doi:https://doi. org/10.1016/j.synthmet.2020.116610.

Ashton, Michael, Nicole Trometer, Kiran Mathew, Jin Suntivich, Christoph Freysoldt, Susan B. Sinnott and Richard G. Hennig. 2019. Predicting the electrochemical synthesis of 2D materials from first principles. The Journal of Physical Chemistry C 123(5): 3180–3187. American Chemical Society. doi:10.1021/acs.jpcc.8b10802.

Carvalho, Wildemar S.P., Menglian Wei, Nduka Ikpo, Yongfeng Gao and Michael J. Serpe. 2018. Polymer-based technologies for sensing applications. doi:10.1021/acs.analchem.7b04751.

Chowdhury, N.K. and B. Bhowmik. 2021. Micro/nanostructured gas sensors: the physics behind the nanostructure growth, sensing and selectivity mechanisms. Nanoscale Advances 3(1): 73–93. RSC. doi:10.1039/D0NA00552E.

Comini, E. 2013. One- and two-dimensional metal oxide nanostructures for chemical sensing. pp. 299–315. *In*: Raivo Jaaniso and Ooi Kiang Tan (eds). Semiconductor Gas Sensors. Woodhead Publishing Series in Electronic and Optical Materials. Woodhead Publishing. doi:https://doi.org/10.1533/978085 7098665.3.299.

Comini, E. and D. Zappa. 2020. One- and two-dimensional metal oxide nanostructures for chemical sensing. pp. 161–184. *In*: Raivo Jaaniso and Ooi Kiang Tan (eds). Semiconductor Gas Sensors, 2nd Ed. Woodhead Publishing Series in Electronic and Optical Materials. Woodhead Publishing. doi:https://doi.org/10.1016/B978-0-08-102559-8.00005-7.

Das, Sagnik, Subhajit Mojumder, Debdulal Saha and Mrinal Pal. 2022. Influence of major parameters on the sensing mechanism of semiconductor metal oxide based chemiresistive gas sensors: a review focused on personalized healthcare. Sensors and Actuators B: Chemical 352: 131066. doi:https://doi. org/10.1016/j.snb.2021.131066.

Deshmukh, Kalim, Tomáš Kovářík and S.K. Khadheer Pasha. 2020. State of the art recent progress in two dimensional mxenes based gas sensors and biosensors: a comprehensive review. Coordination Chemistry Reviews 424: 213514. doi:https://doi.org/10.1016/j.ccr.2020.213514.

Garg, Naini, Akash Deep and Amit L. Sharma. 2021. Metal-organic frameworks based nanostructure platforms for chemo-resistive sensing of gases.

Coordination Chemistry Reviews 445: 214073. doi:https://doi.org/10.1016/j. ccr.2021.214073.

Ghosh, Rajesh, Mohammed Aslam and Hemen Kalita. 2022. Graphene derivatives for chemiresistive gas sensors: a review. Materials Today Communications 30: 103182. doi:https://doi.org/10.1016/j.mtcomm.2022.103182.

Jiao, Long, Joanne Yen Ru Seow, William Scott Skinner, Zhiyong U. Wang and Hai-Long Jiang. 2019. Metal–organic frameworks: structures and functional applications. Materials Today 27: 43–68. doi:https://doi.org/10.1016/j.mattod. 2018.10.038.

Joensen, Per, R.F. Frindt and S. Roy Morrison. 1986. Single-layer MoS$_2$. Materials Research Bulletin 21(4): 457–461. doi:https://doi.org/10.1016/0025-5408(86)90011-5.

Kailasa, Suresh Kumar, Dharaben J. Joshi, Janardhan Reddy Koduru and Naved I. Malek. 2021. Review on mxenes-based nanomaterials for sustainable opportunities in energy storage, sensing and electrocatalytic reactions. Journal of Molecular Liquids 342: 117524. doi:https://doi.org/10.1016/j.molliq. 2021.117524.

Kailasa, Suresh Kumar, Dharaben J. Joshi, Mehul R. Kateshiya, Janardhan Reddy Koduru and Naved I. Malek. 2022. Review on the biomedical and sensing applications of nanomaterial-incorporated hydrogels. Materials Today Chemistry 23: 100746. doi:https://doi.org/10.1016/j.mtchem.2021.100746.

Kausar, Ayesha. 2022. Graphene to Polymer/Graphene Nanocomposites: Emerging Research and Opportunities (Chapter 1: Graphene: structure, properties, preparation, modification, and applications, pp. 1–24). Elsevier. doi:https://doi.org/10.1016/B978-0-323-90937-2.00010-1.

Khazaei, Mohammad, Ahmad Ranjbar, Masao Arai, Taizo Sasaki and Seiji Yunoki. 2017. Electronic properties and applications of mxenes: a theoretical review. Journal of Materials Chemistry C 5(10): 2488–2503. The Royal Society of Chemistry. doi:10.1039/C7TC00140A.

Korotcenkov, G. 2007. Metal oxides for solid-state gas sensors: what determines our choice? Materials Science and Engineering: B 139(1): 1–23. doi:https:// doi.org/10.1016/j.mseb.2007.01.044.

Korotcenkov, G. and B.K. Cho. 2017. Metal oxide composites in conductometric gas sensors: achievements and challenges. Sensors and Actuators B: Chemical 244: 182–210. doi:https://doi.org/10.1016/j.snb.2016.12.117.

Kumar, Rahul, Neeraj Goel, Mirabbos Hojamberdiev and Mahesh Kumar. 2020. Transition metal dichalcogenides-based flexible gas sensors. Sensors and Actuators A: Physical 303: 111875. doi:https://doi.org/10.1016/j.sna.2020.111875.

Lange, Ulrich, Nataliya V. Roznyatovskaya and Vladimir M. Mirsky. 2008. Conducting polymers in chemical sensors and arrays. Analytica Chimica Acta 614(1): 1–26. doi:https://doi.org/10.1016/j.aca.2008.02.068.

Li, Zhijie, Hao Li, Zhonglin Wu, Mingkui Wang, Jingting Luo, Hamdi Torun, PingAn Hu, et al. 2019. Advances in designs and mechanisms of semiconducting metal oxide nanostructures for high-precision gas sensors operated at room temperature. Materials Horizons 6(3): 470–506. The Royal Society of Chemistry. doi:10.1039/C8MH01365A.

Meng, Zheng, Robert M. Stolz, Lukasz Mendecki and Katherine A. Mirica. 2019. Electrically-transduced chemical sensors based on two-dimensional nanomaterials. Chemical Reviews 119(1): 478–598. American Chemical Society. doi:10.1021/acs.chemrev.8b00311.

Mondal, Biplob and Pranjal Kumar Gogoi. 2022. Nanoscale heterostructured materials based on metal oxides for a chemiresistive gas sensor. ACS Applied Electronic Materials 4(1): 59–86. American Chemical Society. doi:10.1021/acsaelm.1c00841.

Naguib, Michael, Olha Mashtalir, Joshua Carle, Volker Presser, Jun Lu, Lars Hultman, Yury Gogotsi, et al. 2012. Two-dimensional transition metal carbides. ACS Nano 6(2): 1322–1331. American Chemical Society. doi:10.1021/nn204153h.

Ping, Jianfeng, Zhanxi Fan, Melinda Sindoro, Yibin Ying and Hua Zhang. 2017. Recent advances in sensing applications of two-dimensional transition metal dichalcogenide nanosheets and their composites. Advanced Functional Materials 27(19): 1605817. doi:https://doi.org/10.1002/adfm.201605817.

Rao, C.N.R., G.U. Kulkarni, P. John Thomas and Peter P. Edwards. 2002. Size-dependent chemistry: properties of nanocrystals. Chemistry—A European Journal 8(1): 28–35. doi:https://doi.org/10.1002/1521-3765(20020104)8:1<28:: AID -CHEM28>3.0.CO;2-B.

Shi, Xinyue, Yuying Shan, Meng Du and Huan Pang. 2021. Synthesis and application of metal-organic framework films. Coordination Chemistry Reviews 444: 214060. doi:https://doi.org/10.1016/j.ccr.2021.214060.

Stock, Norbert and Shyam Biswas. 2012. Synthesis of metal-organic frameworks (MOFs): routes to various mof topologies, morphologies, and composites. Chemical Reviews 112(2): 933–969. American Chemical Society. doi:10.1021/cr200304e.

Verger, Louisiane, Chuan Xu, Varun Natu, Hui-Ming Cheng, Wencai Ren and Michel W. Barsoum. 2019. Overview of the synthesis of mxenes and other ultrathin 2D transition metal carbides and nitrides. Current Opinion in Solid State and Materials Science 23(3): 149–163. doi:https://doi.org/10.1016/j.cossms.2019.02.001.

Wackerlig, Judith and Peter A. Lieberzeit. 2015. Molecularly imprinted polymer nanoparticles in chemical sensing—synthesis, characterisation and application. Sensors and Actuators B: Chemical 207: 144–157. doi:https://doi.org/10.1016/j.snb.2014.09.094.

Wilson, J.A. and A.D. Yoffe. 1969. The transition metal dichalcogenides discussion and interpretation of the observed optical, electrical and structural properties. Advances in Physics 18(73): 193–335. doi:10.1080/00018736900101307.

Xie, Huaguang, Zhong Li, Liang Cheng, Azhar Ali Haidry, Jiaqi Tao, Yi Xu, Kai Xu, et al. 2022. Recent advances in the fabrication of 2D metal oxides. IScience 25(1): 103598. doi:https://doi.org/10.1016/j.isci.2021.103598.

Yaghi, O.M., Guangming Li and Hailian Li. 1995. Selective binding and removal of guests in a microporous metal–organic framework. Nature 378(6558): 703–706. doi:10.1038/378703a0.

Yang, Wei, Lin Gan, Huiqiao Li and Tianyou Zhai. 2016. Two-dimensional layered nanomaterials for gas-sensing applications. Inorganic Chemistry Frontiers 3(4): 433–451. The Royal Society of Chemistry. doi:10.1039/C5QI00251F.

Yoon, Yeosang, Phuoc Loc Truong, Daeho Lee and Seung Hwan Ko. 2021. Metal-oxide nanomaterials synthesis and applications in flexible and wearable sensors. ACS Nanoscience Au 2(2): 64–92. American Chemical Society. doi:10.1021/acsnanoscienceau.1c00029.

Chapter 4

Thin Film Deposition Techniques

4.1 INTRODUCTION

As nanotechnology progresses rapidly, the demand for miniaturized and smart thin films is increasing tremendously. This has led to the development of a new kinds of sophisticated thin-film deposition techniques which provide a sensitive, uniform, flexible, thin film platform for the development of smart chemical sensors (Kumar et al. 2018; Liu and Wang 2020; Zehra et al. 2021). Numerous techniques have been explored for the fabrication of well uniform thin films. Thin film deposition techniques are playing a crucial role in harvesting various properties of materials such as optical, dielectric, electric, etc. The nanomaterial size materials show dissimilar properties compared to bulk, and therefore, their study is needed for various applications. Generally, a thin layer of material is fabricated for thin film coating. The thickness may vary between nanometer to micrometer scale. As the coating size increases, the material properties also get varied. Generally, for effective chemical sensing, nanometer to a few micrometer size is preferred. The thin film structure is divided into two parts, namely, amorphous and crystalline structure. The material structure can be deposited onto the base known as substrate, which can be made up of glass, silica, alumina, etc. However, the selection of substrate is mainly based on the type of mode. Generally, in optical-based applications, the substrate plays a crucial role, and therefore, the choice of substrate is also very important. Another part is a thin layer or sensing layer, which acts as an active region for sensing (Benelmekki and Erbe 2019; Janarthanan et al. 2021).

Generally, the thin film deposition techniques (tabulated in Table 4.1) are divided into two major types mainly depending upon their working principle. First is a physical deposition and other is a chemical deposition and these consist of many sub-techniques as shown below. Figure 4.1 shows various deposition methods used for thin-films fabrication. Here, some of the most efficient and widely adopted techniques for thin-film fabrication are reviewed.

Table 4.1 Various deposition techniques

Physical deposition methods	Chemical deposition methods

1. Evaporation techniques
  (a) Vacuum thermal evaporation
  (b) Electron beam evaporation
  (c) Pulsed laser deposition
  (d) Ion plating
  (e) Arc deposition
  (f) Molecular beam epitaxy
2. Sputtering techniques
  (a) Direct current sputtering
  (b) Radio frequency sputtering
	1. Sol-gel technique
	2. Spray pyrolysis technique
	3. Chemical bath deposition (CBD)
	4. Plating
	5. Chemical vapor deposition (CVD)
	  (a) Low pressure CVD
	  (b) Plasma enhanced CVD
	  (c) Atomic layer deposition

4.2 PHYSICAL DEPOSITION

This is a widely utilized technique for thin film fabrication, in which the thin film is deposited by condensing the vaporized form of the solid material. The physically ejected material gets condensed and nucleated onto the substrate. The typical features of this technique are tabulated in Table 4.2. It consists of the basic steps as follows (Ohring 2002; Paras and Kumar 2021a):

Table 4.2 Typical features of physical vapor deposition. [Reproduced from Ref. (Baptista et al. 2018), Copyright 2018, MDPI]

Parameters	Evaporation	Sputtering
Vacuum	High	Low
Deposition rate	High (up to 750,000 A min^{-1})	Low (except for pure metals and dual magnetron)
Adhesion	Low	High
Absorption	Less absorbed gas into the film	High
Deposited species energy	Low (~0.1–0.5 eV)	High (1–100 eV)
Homogeneous film	Less	More
Grain size	Bigger	Smaller
Atomized particles	Highly directional	More dispersed

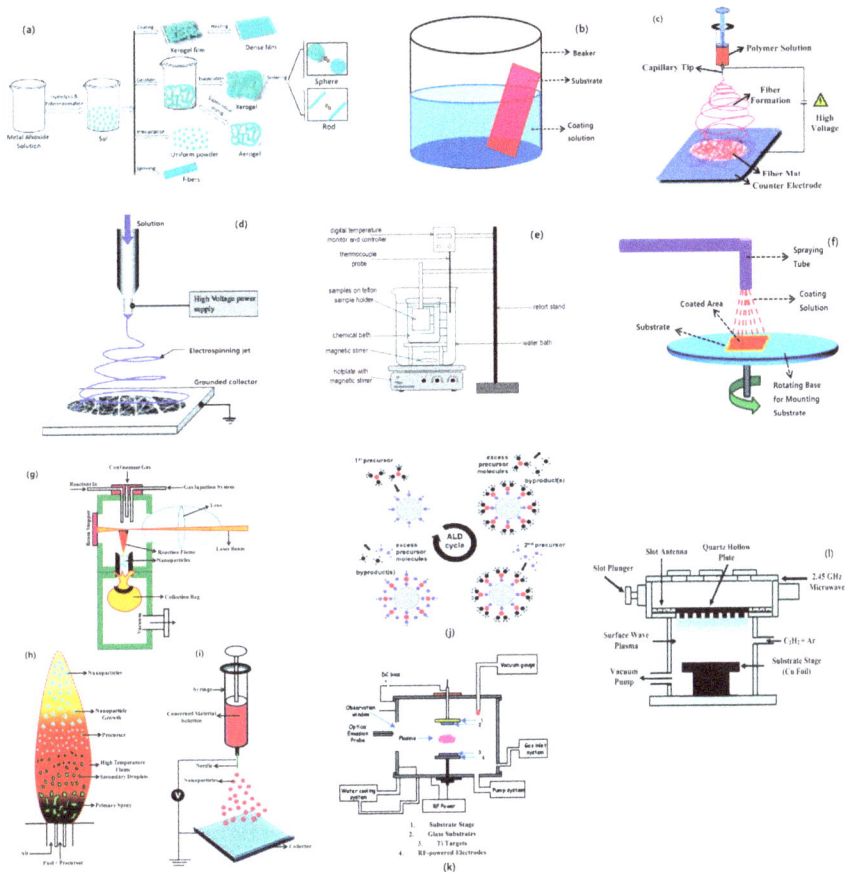

Figure 4.1 Numerous thin-film deposition techniques. (a) Sol–gel processes, (b) Dip coating technique, (c) Conventional electrospinning with vertical feeding of the solution, (d) Electrospinning using a grounded collector, (e) Chemical bath deposition, (f) Spin coating on a substrate, (g) Laser pyrolysis, (h) Flame spray pyrolysis (i) Electrospraying technique, (j) Atomic layer deposition, (k) RF magnetron sputtering, and (l) Microwave-assisted surface-wave plasma chemical vapor deposition technique. [Reproduced from Ref. (Roji et al. 2017) Copyright 2017, Royal Society of Chemistry].

1. The material gets converted into vaporized form by using evaporation, sputtering, or gas form.
2. The deposition of material vapor from material onto the substrate by using a molecular flow, thermal flow, etc. During this process a large number of collisions can occur.
3. Growing the thin-film onto the substrate.

4.2.1   Evaporation Techniques

This is the most common technique and widely adopted due to its simplicity, cost-effectiveness, and ease in depositing materials. The main part it consists of heating the material. As the name suggests, the heat is supplied to the material, which changes the phase of the material and thus, it evaporates and gets condensed onto the substrate. The material should heat up to its melting point so that it will easily evaporate. However, it requires a high vacuum of the order of 10^{-6} Torr, 0.133 mPa, for efficient thin film deposition. The high vacuum assists to increase in the mean free path of vaporized atoms. Here, the mean free path of the depositing atom or molecule is very crucial, and therefore, a long mean free path is necessary. The mean free path shows the average distance of a molecule or atom covered without collision with another molecule, and therefore, not unobstructed (Adeyeye and Shimon 2015; Martín-Palma and Lakhtakia 2013). The mean free path (λ) is given by the following equation (Mabe et al. 2018; Schalk et al. 2022):

$$l = \frac{RT}{\sqrt{2}\pi D^2 N_A P} \qquad (4.1)$$

where, T is the temperature, D is the diameter of a molecule, N_A is the Avogadro's number, P is the pressure, and R is the gas constant. Typically, for a 0.3 nm diameter molecule and at 10^{-4} Pa, the $l = 100$ m is observed. From above equation, one thing is clear that most of the parameters like R, T, D, R, and NA are all constant, and therefore, only at a lower pressure, the mean free path can be extended.

The metals like gold, silver, platinum, indium, titanium, and chromium are mostly deposited by using the evaporation technique. However, certain dielectric materials like SiO_2 thin film preparations have also been reported. For most the metals like (Ag, Au, and Cu), an evaporation temperature of around 1500 K is required. Whereas, at this temperature, materials like Pt and Al_2O_3 hardly get vaporized and thus, require a high temperature due to their low vapor pressure and high melting temperature. The materials can be excited by using different source heating methods like resistive, RF, or electron beam. The detailed procedure is as follows (Janarthanan et al. 2021; Martín-Palma and Lakhtakia 2013; Ohring 2002; Paras and Kumar 2021a):

1. The substrate is placed onto the substrate holder onto which the depositing material gets condensed. The substrate is rotatable so that uniform deposition takes place.
2. First, a small piece of depositing material is placed into the crucible and then heated up to the source material's melting point. Later the material gets condensed onto the substrate.

3. The crucible is a high temperature sustainable material and is made up of carbon, tungsten, Al_2O_3, etc. The resistive support materials are made typically from tungsten and the RF support materials are composites like boron nitride.

4. The shutter plate is used to control the deposition rate.

5. The heating and cooling of the substrate is also provided. The heating controls the condensation of the molecules and cooling regulates the quenching rate of incident species.

6. The thickness monitor is done by using a quartz crystal oscillator which is placed inside the vacuum chamber. The resonating frequency is dependent upon the mass deposited on it.

7. According to the Clausius-Clapeyron equation, the equilibrium vapor pressure at the evaporation surface is dependent upon the temperature.

4.2.2 Sputtering

Sputtering is a process of ejecting atoms from the target or source material due to the bombardment of high energy particles and getting deposited onto the substrate. This is the most preferable method for thin-film preparation due to its ability to deposit metal, alloys and, dielectric materials. This technique is mostly used in semiconductor industries. The thin film fabricated using sputtering has high uniformity and control over thickness.

As ions are charged particles, their velocity, and behavior can be easily controlled by using a magnetic field. The magnets are used behind the target cathode so that the electrons can trap the target and will not bombard the substrate. The typical disposition steps are summarized below (Bandorf et al. 2014; Ganesan et al. 2018; Janarthanan et al. 2021; Kelly and Arnell 2000).

1. In sputtering, the substrate is placed into a high vacuum chamber containing an inert Ar gas at low pressure (1–15 mTorr). Generally, a high molecular weight gas is chosen like Ar or Xenon, which facilitates the high energy collision.

2. The negative charge is applied to the source materials which get deposited onto the substrate by generating the plasma.

3. The free electrons emitted from the target materials react with the Ar, which converts positive charge Ar ions. The positive charge Ar gas gets accelerated towards the negatively charged target with a very high velocity, causing a high energy collision.

4. Each collision leads to ejecting the atoms of the source material with high kinetic energy which is enough to reach up to the substrate.

5. The $E \times B$ drift path is used so that the electron will not follow a straight path and thus, the maximum collision will occur with inert gas. In magnetron sputtering, the electrons are tightly confine in the plasma or near to the target surface.

The sputtering technique is divided into two types namely RF sputtering and DC sputtering.

4.2.2.1 DC sputtering

In direct current sputtering, the target directly conducts the electricity with loss of I^2R. The source materials are bombarded with an ionized gas molecules, which sputters the atoms of the target into the plasma and then these atoms get deposited onto the substrate. This is the best technique for metal coating and is also easy to control the power source. The typical sputter pressure is 0.5 mTorr to 100 mTorr. The DC voltage, usually in the range of –2 to –5 kV, is applied to the target (cathode) and positive biasing applied to the substrate turns out to be the anode. The Ar gas is first ionized and attracted towards the target material and ejects atoms off into the plasma to glow, which gets deposited onto the substrate. The magnets are behind the target so that the electrons will get trapped near the target and not set free to bombard the substrate. This significantly increases the deposition rate over the substrate. While with DC sputtering, only metals can be deposited efficiently (Bellardita et al. 2019; Ganesan et al. 2018; Janarthanan et al. 2021). Also, an accumulation of charge on the target material may lead to hampering its use. The arcing creates a problem in which the intensely focused and localized discharge emit from the source material leads to non-uniform film thickness and thus, raises quality issues.

4.2.2.2 RF magnetron sputtering

RF magnetron sputtering is a high vacuum sputtering method (Figure 4.1k) capable to coat both metals and dielectric materials. In this, an alternating electrical potential of the current at radio frequency is used to solve the problem of arcing or charge building on the target material. During each cycle, the charge will build up on the target material. During the positive cycle, the electrons will attract toward the target making a negative bias. While in a negative cycle, which occurs at a radio frequency of 13.56 MHz, ion bombardment to the target and sputtering take place continuously. Typical RF frequencies between 5–30 MHz can be used but mostly the 13.56 MHz is widely adopted. In RF sputtering, the target is always cleaned off the plasma which can sustain throughout the chamber with a lower pressure of 1–15 mTorr. In the RF sputtering, the magnetic field traps the electron near the target and hence, increase the deposition rate (Bosco et al. 2012; Madhuri 2020; Mattox 2010). While in RF frequency, the rate of deposition is slower than the DC sputtering. Generally, it takes 1012 V to sputter dielectric materials. This technique is also used to coat insulating materials.

4.3 CHEMICAL DEPOSITION

The chemical deposition technique is widely adopted due to its simple, cost-effectiveness, and high efficiency. The mass-production of thin film fabrication is easily achieved by using this technique. The chemical vapor deposition, chemical bath deposition, sol–gel method, electro-spraying technique, atomic layer deposition, and laser pyrolysis are the most utilized techniques for thin-film fabrication. A detailed description of these methods is discussed below.

4.3.1 Sol-gel Method

This is the most applied technique (Figure 4.1a) to a prepare good quality thin-film of ceramic materials like metal oxides, nitrides, and carbides. Different kinds of porous structures can be easily synthesized by this process. Typical sol-gel method consists of a metal alkoxides, water acting as hydrolysis agent, alcohol acting as solvent, and acid or base acting as a catalyst. In the first step, a precursor is dissolved in a solvent solution. The acid or base is used as a catalyst to control the condensation and stabilize the sol. The properties are mainly dependent upon the rate of condensation and hydrolysis. With a slower hydrolysis rate, the small size of nanoparticles can be obtained. The shape is formed during sol to gel transformation. Heating at a certain temperature of around 100 °C will remove the water and other solvents and will densify the metal-oxide structure.

For a thin-film preparation, the typical sol-gel dip coating or spin coating technique is preferred. In this technique, there are two methods for synthesis (Brinker and Scherer 1990; Paras and Kumar 2021b; Scriven 1988):

1. Colloidal route: water is used as a solvent in the precursor preparation.
2. Polymeric rote: alcohol acts as a solvent for the preparation of precursor.

The following steps are involved while making a sol-gel thin film:

1. The substrate such as glass, silicon, and ITO, can be used. The substrate should have good wetting and adhesive properties.

2. The film thickness ranging from 20 nm to 500 nm can be prepared via a dip coating or spin coating. In a dip coating, the rate of deposition depends upon the withdrawal speed while in the spin coating, the rate is mainly dependent upon the rotation, viscosity of the gel, and time. A thick film of over 500 nm is not preferred as it may generate cracks during the sintering process.

3. In spin coating, the sol-gel is placed onto the substrate and then it is allowed to rotate at a certain speed. The thickness of the film depends on the viscosity, rotary speed, ratio of the solution, and the solvent used.

4. Then controlled heat treatment under a controlled humidity environment is applied so that the solvent will evaporate and the film becomes densified.

4.3.2   Electrospinning Method

Electrospinning principle was first proposed by Raleigh in 1897 and patented by Formhals in 1934. It is a spinning technique (Figure 4.1c, d) in which electrostatic principle is used to produce thin fibers of diameter nm to micrometer range. By using this method, almost all types of polymers including synthetic polymers, natural polymers, or a blend can be coated very easily. The working steps are as follows (Davoodi et al. 2021; Ksapabutr and Panapoy 2022; Patel et al. 2021):

1. It contains a high voltage power supply, a spinneret, and a grounded collector plate.

2. A polymeric solution is first prepared by dissolving the polymer into a solvent and then filled into the capillary tube for injection.

3. A high voltage source is used to inject the solution onto the collector plate. For this purpose, one end of the power supply is connected to the tip of a capillary tube and another end to the conductor plate.

4. The liquid undergoes under surface tension in the capillary tube. The electric field induces surface charge to the liquid surface and when it reaches the critical value then repulsive electrical force becomes greater than the surface tension of the liquid, which results in the charge jet formation between the capillary tip and the collector.

5. The solution is ejected and the polymer is left behind.

4.3.3 Chemical Bath Deposition

This is the most common technique (Figure 4.1e) used for making an inorganic and non-metallic thin film on the substrate. The thin-film preparation steps are as follows (Jilani 2017; Pawar et al. 2011; Koao et al. 2014).

1. In a typical method, the bath solution is prepared by using one or two metal salts (M) (such as chlorides, nitrates, sulfates, or acetates) and the chalcogenide (X) (like O, S, Se) in an aqueous liquid.
2. A complexing agent may be added to control the growth of a solid. The complexing agents' own greater affinity towards metals and thus, provides ligands for the metals.
3. The dipping time varies ranging from 6 to 72 hours and also the temperature of the reaction.
4. The role of pH, temperature, and concentration is very important in CBD. As the concentration of metal salt increases it corresponds to an increase in the reaction speed. The increase in temperature increases the supply of hydroxyl ions and also the deprotonation of hydronated metal species. Thus, it increases the formation of a solid.

It consists of three steps:

1. Formation of atomic/molecular/ionic species
2. Transportation of species across a channel
3. Condensation of species

These dissolved species connect to the substrate as per the ion-by-ion process. This directly develops nuclei of inorganic film on the substrate by heterogeneous nucleation. Due to hydrolysis and condensation reactions, the colloidal particles might produce in the solution further and get attracted towards the substrate via Vander walls or electrostatic attraction. This may lead to formation of a polycrystalline film. For non-oxide films, the source of chalcogenide anions can be controlled easily by adjusting the pH, temperature, and concentration of the chalcogenide source. For oxide films, hydrolysis is important, and therefore, the pH and temperature are very crucial to control the film properties.

4.3.4 Spray Pyrolysis Method

In this method, the solution is sprayed on the heated surface which leads to the formation of a chemical layer due to their reactions occured on the surface (Figure 4.1h). This is relatively a simpler and cost-effective technique for thin film preparation. Both types of thin and thick films can be easily prepared by using this method.

The main principle is based on aerosols that get deposited onto the heated substrate. The following key steps are involved in this method (Falcony et al. 2018; Janarthanan et al. 2021).

1. It contains an atomizer, precursor solution, substrate heating, and temperature controller. The atomizer can be of different types such as air blast, ultrasonic, and electrostatic.

2. The solution is atomized into small droplets and then it is directed towards the heated substrate which decomposes the precursor to produce a thin film over the substrate.

3. The deposition process consists of four steps:
   (a) A solid film can form by attaching the droplets to the heated substrate where it gets dried,
   (b) A dried solid gets deposited on the heated substrate and decompose further. The solvent gets evaporated before it reaches the substrate,
   (c) The solvent gets vaporized as it approaches towards the heated base, the solid gets deposited and evaporates and the vapor disperses over the substrate and
   (d) Complete process occurs in the vapor state. Generally, most commonly occurring process is in type (a) and type (b).

4. In spray-pyrolysis method, certain factors are important, for example, rate of gas current, jet to substrate gap, surrounding temperature, drop size, solution concentration and stream, transporter gas, and temperature of the substrate.

5. General parameters like heater power: 1 kW, flow speed: 20 L/min, and pressure: 20 psi are used. An inert gas can also be used if any chance of contamination occurs.

Certain parameters that are required to be considered:

1. The temperature is the main parameter in spray-pyrolysis method. Temperature controls the morphology and properties of the material. As temperature increases, the film's nature changes from a crack or dense to a porous structure. It also changes the electrical and optical properties of the materials.

2. The composition of precursor solution can change the properties of the deposited film. Generally, the grain size rises with a higher concentration of precursor in ethanol solvent.

3. The pH of the solution also influences the growth rate. The rate is more significant in $3.5 \leq \text{pH} \leq 4.3$. At a low pH, the growth rate seems to decrease.

4.3.5   Chemical Vapor Deposition

This method mainly does film deposition via the chemical reaction and surface adsorption (Figure 4.1(l)). In the reaction chamber, the reactant gases react and condense into non-volatile film onto the substrate. This method is widely explored for amorphous and polycrystalline thin layers. The system consists of

1. feed lines for gases;
2. mass flow regulators for gases
3. a reaction chamber or reactor;
4. a substrate heater; and
5. temperature sensors.

It consists of the following major steps (Choy 2003; Dahmen 2003; Teixeira et al. 2011).

1. First, a controlled mixture of reactant gases and diluted gases is introduced to the reaction chamber. The gas species flow towards the substrate and the reactant gas species deposit onto the substrate.

2. The chemical reactions occur on the substrate and the film gets deposited.

3. The gaseous bi-product is evacuated from the reaction chamber.

4. In this method, the reaction occurs with the substrate surface known as heterogeneous reaction. These are strongly adhered to the substrate surface and generate a good quality film. Whereas, the reactions which occur in the gas phase are known as homogeneous reactions. This generates a poor-quality film.

Based on the operating pressure range, CVD is classified into different types (Table 4.3) as follows (Benelmekki and Erbe 2019; Choy 2003).

1. **Atmospheric pressure CVD (APCVD):** It operates at an atmospheric level.

2. **Low-pressure CVD (LPCVD):** The reactor operates at an average vacuum (30–250 Pa). The temperature (600–900 °C) is higher than APCVD.

3. **Plasma enhanced CVD (PECVD):** It uses high frequency waves (microwave: 2.45 GHz, ultrahigh or radio frequencies: 0~13.56 MHz) for plasma induction. It operates at low pressure and temperature (< 900 °C) and uses RF produced glow ejection, which moves energy to reactant gases.

Table 4.3 Advantages and limitations of various CVD-based methods

Type	Advantages	Disadvantages
APCVD	Simple, Fast Deposition,	Poor Step Coverage, Contamination
LPCVD	Low Temperature Excellent Purity, Excellent Uniformity, Good Step Coverage, Large Wafer Capacity	High Temperature, Slow Deposition
PECVD	Low Temperature, Good Step Coverage	Chemical and Particle Contamination

4.4 CONCLUSION AND OUTLOOK

The current demands of thin films for chemical sensing and other applications are tremendously growing, and therefore, flexible design, efficient, simple, less time consuming, low-cost thin film deposition and coating techniques are highly required. This chapter reviews a few most widely utilized thin film deposition techniques for the detection of chemical and biochemical sensing. Among the two main categories, physical deposition is mainly divided into evaporation and sputtering techniques and another side the chemical deposition is categorized into different methods like sol-gel, chemical bath deposition, spray pyrolysis, and chemical vapor deposition. Considering the extensive demand for thin films for numerous environmental and industrial applications, this chapter provides a precise discussion of the physics and chemical principles behind these deposition techniques.

REFERENCES

Adeyeye, A.O. and G. Shimon, 2015. Growth and characterization of magnetic thin film and nanostructures. pp. 1–41. *In*: R.E. Camley, Z. Celinski and R.L. Stamps (eds). Magnetism of Surfaces, Interfaces, and Nanoscale Materials. Handbook of Surface Science Stamps Vol. 5., North-Holland. Elsevier. doi:https://doi.org/10.1016/B978-0-444-62634-9.00001-1.

Bandorf, R., V. Sittinger and G. Bräuer, 2014. High power impulse magnetron sputtering—HIPIMS. pp. 75–99. *In*: Saleem Hashmi, Gilmar Ferreira Batalha, Chester J. Van Tyne and B.T. Bekir (eds). Comprehensive materials processing. Vol. 4: Films and coatings. Technology and recent development. Oxford: Elsevier. doi:https://doi.org/10.1016/B978-0-08-096532-1.00404-0.

Baptista, Andresa, Francisco Silva, Jacobo Porteiro, José Míguez and Gustavo Pinto. 2018. Sputtering physical vapour deposition (PVD) coatings: a critical review on process improvement and market trend demands. Coatings 8(11): 402. https://doi.org/10.3390/coatings8110402.

Bellardita, Marianna, Agatino Di Paola, Sedat Yurdakal and Leonardo Palmisano. 2019. Preparation of catalysts and photocatalysts used for similar processes. pp. 25–56. *In*: G. Marcì and L. Palmisano (eds). Heterogeneous Photocatalysis. Elsevier. doi:https://doi.org/10.1016/B978-0-444-64015-4.00002-X.

Benelmekki, Maria and Andreas Erbe. 2019. Frontiers of Nanoscience. (Chapter 1: Nanostructured thin films–background, preparation and relation to the technological revolution of the 21st century. pp. 1–34). Elsevier. doi:https://doi.org/10.1016/B978-0-08-102572-7.00001-5.

Bosco, Ruggero, Jeroen Van Den Beucken, Sander Leeuwenburgh and John Jansen. 2012. Surface engineering for bone implants: a trend from passive to active surfaces. Coatings 2(3): 95–119. doi:10.3390/coatings2030095.

Brinker, C. Jeffrey and George W. Scherer. 1990. Sol-Gel Science: The Physics and Chemistry of Sol-Gel Processing. (Chapter 4: Particulate sols and gels. pp. 234–301). San Diego: Academic Press. doi:https://doi.org/10.1016/B978-0-08-057103-4.50009-X.

Choy, K.L. 2003. Chemical vapour deposition of coatings. Progress in Materials Science 48(2): 57–170. doi:https://doi.org/10.1016/S0079-6425(01)00009-3.

Dahmen, Klaus-Hermann. 2003. Chemical vapor deposition. pp. 787–808. *In*: Robert A. Meyers (ed.). Encyclopedia of Physical Science and Technology, 3rd Ed. Academic Press. doi:https://doi.org/10.1016/B0-12-227410-5/00102-2.

Davoodi, Pooya, Elisabeth L. Gill, Wenyu Wang and Yan Yan Shery Huang. 2021. Advances and innovations in electrospinning technology. pp. 45–81. *In*: Naresh Kasoju and Hua Ye (eds). Biomedical Applications of Electrospinning and Electrospraying. Woodhead Publishing. doi:https://doi.org/10.1016/B978-0-12-822476-2.00004-2.

Falcony, Ciro, Miguel A. Aguilar-Frutis, and Manuel García-Hipólito. 2018. Spray pyrolysis technique; high-K dielectric films and luminescent materials: a review. Micromachines 9(8): 414. doi:10.3390/mi9080414.

Ganesan, R., B. Akhavan, X. Dong, D.R. McKenzie and M.M.M. Bilek. 2018. External magnetic field increases both plasma generation and deposition rate in HiPIMS. Surface and Coatings Technology 352: 671–679. doi:https://doi.org/10.1016/j.surfcoat.2018.02.076.

Janarthanan, B., C. Thirunavukkarasu, S. Maruthamuthu, M. Aslam Manthrammel, M., Md. Shkir, S. AlFaify, M. Selvakumar, et al. 2021. Basic deposition methods of thin films. Journal of Molecular Structure 1241: 130606. doi:https://doi.org/10.1016/j.molstruc.2021.130606.

Jilani, Asim, Md S. Abdel-wahab and A.H. Hammad. 2017. Advance deposition techniques for thin film and coating. pp. 1–14. *In*: N.N. Nikitenkov (ed.). Modern Technologies for Creating the Thin-film Systems and Coatings. IntechOpen. doi:10.5772/65702.

Kelly, P.J. and R.D. Arnell. 2000. Magnetron sputtering: a review of recent developments and applications. Vacuum 56(3): 159–172. doi:https://doi.org/10.1016/S0042-207X(99)00189-X.

Koao, L.F., F.B. Dejene, and H.C. Swart. 2014. Properties of flower-like ZnO nanostructures synthesized using the chemical bath deposition. Materials Science in Semiconductor Processing 27: 33–40. doi:https://doi.org/10.1016/j.mssp.2014.06.009.

Ksapabutr, Bussarin and Manop Panapoy. 2022. Fundamentals of electrospinning and safety. pp. 3–30. *In*: Vincenzo Esposito and Debora Marani (eds). Metal Oxide-Based Nanofibers and Their Applications—A volume in Metal Oxides. Elsevier. doi:https://doi.org/10.1016/B978-0-12-820629-4.00004-7.

Kumar, Pawan, Ki-Hyun Kim, Kowsalya Vellingiri, Pallabi Samaddar, Parveen Kumar, Akash Deep and Naresh Kumar. 2018. Hybrid porous thin films: opportunities and challenges for sensing applications. Biosensors and Bioelectronics 104: 120–137. doi:https://doi.org/10.1016/j.bios.2018.01.006.

Liu, Wenlong and Hong Wang. 2020. Flexible oxide epitaxial thin films for wearable electronics: fabrication, physical properties, and applications. Journal of Materiomics 6(2): 385–396. doi:https://doi.org/10.1016/j.jmat.2019.12.006.

Mabe, Taylor L., James G. Ryan and Jianjun Wei. 2018. Functional thin films and nanostructures for sensors. pp. 169–213. *In*: Ahmed Barhoum and Abdel Salam (eds). Fundamentals of nanoparticles. Classifications, Synthesis Methods, Properties. Elsevier. doi:https://doi.org/10.1016/B978-0-323-51255-8.00007-0..

Madhuri, K.V. 2020. Thermal protection coatings of metal oxide powders. pp. 209–231. *In*: Yarub Al-Douri (ed.). Metal Oxide Powder Technologies: Fundamentals, Processing Methods and Applications. Elsevier. doi:https://doi.org/10.1016/B978-0-12-817505-7.00010-5.

Martín-Palma, Raúl J. and Akhlesh Lakhtakia. 2013. Vapor-deposition techniques. pp. 383–398. *In*: Akhlesh Lakhtakia and Raúl J. Martín-Palma (eds). Engineered Biomimicry. Boston: Elsevier. doi:https://doi.org/10.1016/B978-0-12-415995-2. 00015-5.

Mattox, Donald M. 2010. Handbook of Physical Vapor Deposition (PVD) Processing, 2nd Ed. (Chapter 5: The low pressure plasma processing environment, pp. 157–193). Willam Andrew: Applied science publishers. doi:https://doi.org/10.1016/B978-0-8155-2037-5.00005-8.

Ohring, Milton. 2002. Materials Science of Thin Films: Deposition and Structures, 2nd Ed. (Chapter 3: Thin-film evaporation processes, pp. 95–144). San Diego: Academic Press. doi:https://doi.org/10.1016/B978-012524975-1/50006-9.

Paras and Aditya Kumar. 2021a. Anti-bacterial and anti-viral polymeric coatings. pp. 776–785. *In*: M.S.J. Hashmi (ed.). Encyclopedia of Materials: Plastics and Polymers, Vol. 2. Elsevier. https://doi.org/10.1016/B978-0-12-820352-1.00118-8.

Paras and Aditya Kumar. 2021b. Anti-wetting polymeric coatings. pp. 786–795. *In*: M.S.J. Hashmi (ed.). Encyclopedia of Materials: Plastics and Polymers, Vol. 2. Elsevier. doi:https://doi.org/10.1016/B978-0-12-820352-1.00141-3.

Patel, Kapil D., A.R. Padalhin, Rose A.G. Franco, Fiona Verisqa, Hae Won Kim and Linh Nguyen. 2021. Basic concepts and fundamental insights into electrospinning. pp. 3–43. *In*: Naresh Kasoju and Hua Ye (eds). Biomedical Applications of Electrospinning and Electrospraying. Woodhead Publishing. doi:https://doi.org/10.1016/B978-0-12-822476-2.00010-8.

Pawar, S.M., B.S. Pawar, J.H. Kim, Oh-Shim Joo and C.D. Lokhande. 2011. Recent status of chemical bath deposited metal chalcogenide and metal oxide thin films. Current Applied Physics 11(2): 117–161. doi:https://doi.org/10.1016/j.cap.2010.07.007.

Roji M., Ani Melfa, Jiji G. and Ajith Bosco Raj T. 2017. A retrospect on the role of piezoelectric nanogenerators in the development of the green world. RSC Advances, The Royal Society of Chemistry 7(53): 33642–33670. doi:10.1039/C7RA05256A.

Schalk, Nina, Michael Tkadletz and Christian Mitterer. 2022. Hard coatings for cutting applications: physical vs. chemical vapor deposition and future challenges for the coatings community. Surface and Coatings Technology 429: 127949. doi:https://doi.org/10.1016/j.surfcoat.2021.127949.

Scriven, L.E. 1988. Physics and applications of dip coating and spin coating. MRS Online Proceedings Library 121(1): 717–729. doi:10.1557/PROC-121-717.

Teixeira, V., J. Carneiro, P. Carvalho, E. Silva, S. Azevedo and C. Batista. 2011. High barrier plastics using nanoscale inorganic films. pp. 285–315. *In*: José-María Lagarón (ed.). Multifunctional and Nanoreinforced Polymers for Food Packaging. Woodhead Publishing. doi:https://doi.org/10.1533/9780857092786.1.285.

Zehra, Nehal, Laxmi Raman Adil, Arvin Sain Tanwar, Subrata Mondal and Parameswar Krishnan Iyer. 2021. Thin-film devices for chemical, biological, and diagnostic applications. pp. 369–405. *In*: Soumen Das and Sandip Dhara (eds). Chemical Solution Synthesis for Materials Design and Thin Film Device Applications. Elsevier. doi:https://doi.org/10.1016/B978-0-12-819718-9.00020-0.

Chapter 5

Gas Sensor

5.1 INTRODUCTION

Climate change and its impact on every living organism on the earth is one of the key challenges at present. The uncontrolled and excessive release of toxic gases and volatile organic compounds into the earth's atmosphere is considered the main reason behind this negative impact. The primary sources of air pollution include vehicles, cars, chemical industrial plants, oil refineries, and natural sources, such as volcanos, wildfires, dust, etc. The air composition has been changed due to different gases discharged from these sources (Patel et al. 2021). As per the WHO report, around seven million people die worldwide because of air pollution. The requirements like the balance of environment and human safety are very important at this stage. Therefore, the real-time detection of different toxic gases and hazardous chemicals is highly required.

Since the last decade, rapid development has been observed in the development of versatile gas sensing devices for monitoring on various toxic gases. Gas sensors are capable of detecting gas molecules from the environment and can ensure the safety of living being at an early stage. The first gas sensor was reported in 1816, known as a Davy lamp, and consisted of an oil flame covered with a mesh screen (Fuśnik et al. 2022). After that, tremendous advancements in the design and development of new sensing methods, such as chemiresistive, electrochemical, fiber optics, surface plasmon resonance (SPR), photonics chip, capacitive type, and quartz crystal microbalance have been realized for highly sensitive and selective chemical and biochemical sensing.

For the development of gas sensors, various nanomaterials such as metal oxides, 2D materials, and polymers are extensively used

(Nasri et al. 2021; Zhang et al. 2022a). These materials' physical, chemical, optical, and electrical properties have been greatly investigated for the effective detection of different types of gases.

This chapter addresses the gas sensor development based on numerous methods, such as electrochemical, capacitive, chemiresistive, field effect transistor, fiber optics, and quartz crystal microbalance. The details such as the advantages and limitations of each method are highlighted.

5.2 DIFFERENT TYPES OF GAS SENSING METHODS

5.2.1 Electrochemical Gas Sensors

The exfoliated graphite (EEG) using KIO_3 as the electrolyte based electrochemical sensor was used to detect NO_2 gas (Łukowiec et al. 2022). Specifically, the developed sensor consisting of EEG and Ag@EEG was utilized for NO_2 sensing at different temperature. It was verified that at higher temperature and NO_2 concentration, the conductivity changed from p-type to n-type. The loading of Ag nanoparticles into the graphite was controlled and detected that at a lower loading concentration of 1.56%, the sensor showed better time characteristics. The electrochemical sensor was developed for gas detection. Firstly, the model was developed and correspondingly the experiment was done (SebtAhmadi et al. 2021). The PTFE-Carbon materials-based sensors with different membranes with different porosity and thicknesses were fabricated as per the model. The designed model and the experimental results were properly validated. The sensor was tested to CO in the varied concentration range of 1–3%. It was observed that for a sensitive sensor design, the polymer should have larger pores. The measurement of gas concentration in an inert atmosphere is also important for environmental applications. The proton conducting material like $La_{0.9}Sr_{0.1}YO_{3-\delta}$ was used as a proton current and successfully employed for determining CO_2 concentration between 2% to 14 vol% at an elevated temperature range from 500–600 °C (Kalyakin et al. 2021). However, this device exhibited a very high response time of about 30 min.

At present, specific research has been focused to increase the tenability and robust nature of an electrochemical sensor; the ionic liquids (IL) are very promising for such applications (Doblinger et al. 2022). However, only bare ionic liquid could initiate some issues like leakage or flow from the sensory part, and therefore, it can mix with some polymer to acquire some stability. In this domain, the IL/poly(IL) with a certain ratios were mixed and employed for gas sensing. The structure provided the response of –2.1 nA/ppm with a low detection limit of a few ppm to SO_2. An electrochemical impedance spectroscopy (EIS) technique used for gas sensing is displayed in Figure 5.1. Generally, it consists of three types of electrode systems, such as working electrode (WE), counter electrode (CE), and reference electrode (RE), as showed in Figure 5.1(a). Gas sensor fabricated using two electrodes system is showed in Figures 5.1(b), (c). The interface between the sensing material and electrode material was effectively used for the detection of vapor molecules. The yttrium stabilized zirconia (YSZ) electrolyte and an $La_{0.8}Sr_{0.2}MnO_3$ (LSM) based electrodes were mixed in such way that their interface played a crucial role in ethanol sensing (Cheng et al. 2021a). At 500 °C, the materials showed a response of –1.26 mv/decade toward 50 ppb –1 ppm of ethyl alcohol. The sensor stability, humidity effect, and repeatability were also tested. However, the sensor working temperature is quite high.

A summary of a few electrochemical-based gas sensors using different materials is tabulated in Table 5.1.

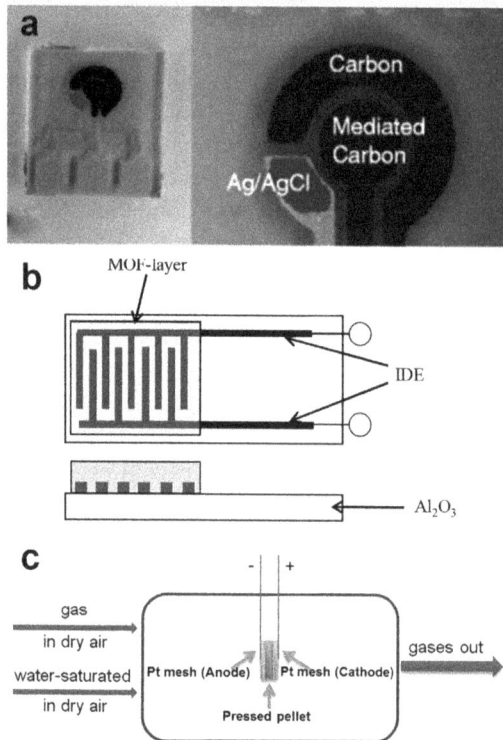

Figure 5.1 (a) Co-phthalocyanine used screen-printed three electrodes for electrochemical impedance spectroscopy (EIS), (b) Top sight of metal organic framework (MOF) paste screen-printed on laser-patterned interdigital electrodes (IDE), and (c) Graphic of gas-sensing device based on two electrodes MOF pellets. [Reproduced with permission (Yao et al. 2021). Copyright 2021, Elsevier].

Table 5.1 Gas-sensing properties of various materials-based electrochemical gas sensors

Materials used	Gas analyte	T (°C)	Detection range	Response/ Recovery time (s)	References
Ag@Exfoliated graphite	NO_2	RT	20–500 ppm	810/2664	(Łukowiec et al. 2022)
Sniffer4D device	NO_2	10.8–41.2	0.5–125 ppb	–	(Liang et al. 2021)
PTFE-Carbon	CO	–	1–3%	140/140	(SebtAhmadi et al. 2021)
Sniffer4D device	CO	10.8–41.2	288–2788 ppb	–	(Liang et al. 2021)
$AgNbO_3$	NH_3	400	50–400 ppm	–	(Li et al. 2021)
IL/poly (IL) 60/40 wt%	NH_3	–	20–500 ppm	–	(Doblinger et al. 2022)
$La_{0.9}Sr_{0.1}YO_{3-\delta}$	CO_2	500–600	~2–14 vol%	~30 min	(Kalyakin et al. 2021)
Ionic liquid/poly (IL) (60/40 wt%)	SO_2	–	25–500 ppm	–	(Doblinger et al. 2022)
$rGO/Fe_3O_4/Cu_2O$	H_2S	RT	500 pM–100 µM	–	(Gu et al. 2021)
Sniffer4D device	O_3	10.8–41.2	0.5–138 ppb	–	(Liang et al. 2021)
Ionic liquid/poly (IL) (60/40 wt%)	O_2	–	20–100 vol%	–	(Doblinger et al. 2022)
TCN(II)– KOH–rGO/CF	N_2O	RT	1–16 ppm	–	(Ramu et al. 2021)
LIG/Co–Pt/ Nafion/PDM	DO	–	30 µM–400 µM	2	(Faruk Hossain et al. 2021)
Glassy carbon electrode/ CoTRP/4GO	DO	RT	0–1.33 mM	–	(Saravia et al. 2016)
Colloidal graphene	LPG	RT	50–800 ppm	–18/–20	(Shukla et al. 2021)
Ag_2O NPs/Au	4-Nitrotoluene	–	0.6–5.9 µM; 37–175 µM	–	(Chakraborty et al. 2022)
NH_2-Fe-MIL-88B/ OMC-3	p-nitrotoluene	–	20–225 µM; 225–2600 µM	–	(Yuan et al. 2018)
Glassy carbon electrode/α-MNO_2	p-nitrotoluene	–	0.2–0.8 µM	–	(Ahmad et al. 2017)
ZnS/Au/f-MWCNT	4-nitrophenol	RT	10–150 µM	–	(Naik et al. 2021)
TCN(II)/KOH/Ni foam	Chlorobenzene	25	1–50 ppm	250	(Silambarasan and Moon 2021)
$La_{0.8}Sr_{0.2}MnO_3$	Ethanol	500	50 ppb–1 ppm	40/52	(Cheng et al. 2021a)
$Bi_{0.5}La_{0.5}FeO_3$	Acetone	580	1–50 ppm	–	(Liu et al. 2019b)

5.2.2 Capacitive-based Gas Sensors

The capacitive-based gas sensing has attracted considerable attention due to its flexibility, simple structure, and ability to detect any type of gas moieties. Table 5.2 shows the recent development of capacitive-based gas sensors.

Table 5.2 Summary of a few capacitive-based gas sensors

Materials used	Gas analyte	Concentration (ppm)	Sensor response	Response/ Recovery time (s)	References
Fum-fcu-MOF	H_2S	100	~0.45%	–	(Yassine et al. 2016)
Ag_2O/ UiO–66(Zr)/NO_2	H_2S	100	~90%	3.33 min	(Surya et al. 2019)
Graphene oxide/ polyaniline	NH_3	100	49.3×10^{-5}/ppm	200	(Wu et al. 2021)
$Mn_{0.5}Zn_{0.5}Fe_2O_4$	NH_3	14	29.3%	3/3–5	(Paul and Philip 2022)
MFM-300 (In)	SO_2	1	0.0016%	–	(Chernikova et al. 2018)
Pd@GaN/Si	H_2	1000	7.88%	1.3	(Yu et al. 2021)
$Mn_{0.5}Zn_{0.5}Fe_2O_4$	Acetone	14	0.3%	–	(Paul and Philip 2022)
$Mn_{0.5}Zn_{0.5}Fe_2O_4$	Ethanol	14	0.2%	–	(Paul and Philip 2022)
MIL-96 (Al)	Methanol	5000	~5%	10–15 min 10–15 min	(Andrés et al. 2020)
MOF-74 (Mg)	Benzene	100	~0.15%	–	(Yuan et al. 2019)
$Cu(BDC)(H_2O)_x$	Toluene	500	0.1%	–	(Sapsanis et al. 2015)

$$R = \frac{C_g - C_0}{C_0} \ \text{ or } \ R = \frac{C_g - C_0}{C_0} \ \text{ or } \ R = \frac{C_g - C_0}{C_0} \times 100$$

The capacitive interdigital electrode structures have been extensively used for the recognition of gases and vapors. For instance, a fumarate-based fcu-MOF (fum-fcu-MOF) structure worked as a sensitive layer and was employed for H_2S detection (Yassine et al. 2016). Their platform consisted of a silicon wafer on the Au coated by PVD method and the electrode patterning was fabricated by a lithography technique. The structure successfully measured the H_2S concentration from 1–100 ppm with an obtained a response of 0.45% and very low limit of detection (LOD) of 5.4 ppb. The Metal oxides (MOXs) and polymer nanocomposite were also employed for gas sensing applications. For instance, the graphene oxide/polyaniline (GO/PANI) composites were tested for NH_3 sensing (Wu et al. 2021). This device exhibited a large sensitivity (49.3×10^{-5}/ppm) and a fast response of ~200 s to NH_3 concentration of 0–100 ppm. The sensing mechanism is based on the p-n type heterojunction, in which the permittivity of the material was changed with the adsorption of NH_3 molecules. This structure realized a flexible device that could bend, and therefore, can be coated to different structures like pipes, storage tanks, and other curved spaces. The metal nanoparticles were also effectively used in capacitive-based sensors. In detail, a Schottky junction was formed between the Pd-GaN, which caused a significant change in the conductivity, and therefore, significant change in the capacitance of the material observed (Yu et al. 2021). The catalytic properties of the metal Pd effectively dissociated the H_2 molecules, and therefore, greatly diffused onto the Pd/GaN Schottky interface, which changed the capacitance of the Schottky junction. This generated a response of 7.88% with a fast response time of 1.3 s to 1000 ppm of H_2 and low LOD at 70 °C. Another IDE-based capacitive sensor fabricated using a MIL-96(Al) MOF thin film on Si/SiO_2 substrate formed using a Langmuir–Blodgett (LB) technique over IDE chip was reported (Andrés et al. 2020). Response of the device was measured for monolayer thin film and drop casted thin film. The change in capacitance was larger for water followed by methanol vapor. The potential use of the LB technique was highlighted.

5.2.3 Chemiresistive Approach

5.2.3.1 Only Metal Oxides Used Room Temperature Gas Sensors

At present, the development of room temperature gas sensors has become a very attractive field to various researchers due to their enormous advantages, such as low power consumption, low-cost, and flexibility at lower temperature. The following Table 5.3 shows the development of gas sensors based on n-type and p-type semiconducting materials.

A room temperature NO_2 sensor is constructed on SnO_2 nanocrystals synthesized by chemical precipitation (Wei et al. 2016). The results declared that the SnO_2 nanocrystals annealed in a vacuum exhibited a much enhanced gas sensing response than SnO_2 annealed in air. The sensor showed a response of around 33 to 11 ppm of NO_2 but exhibited a little longer regaining time of 250 s. Also, the sensor was not studied with other gases to verify its cross-selectivity. The H_2S gas sensor was also fabricated by using various nanomaterials. The ZnO nanorods were used to detect H_2S gas concentration (Hosseini et al. 2015).

Table 5.3 Summary of a few room temperature gas sensing properties of n-type and p-type semiconducting MOXs

Materials used	Gas analyte	Concen- tration (ppm)	Sensor response	Response/ Recovery time (s)	References
WO_3	NO_2	150	392%	32/39	(Khudadad et al. 2021)
SnO_2	NO_2	11	33	100/250	(Wei et al. 2016)
WO_3	NO_2	0.3	15.1	670/2940	(Zhao et al. 2020)
CO_3O_4	NO_2	60	22%	25/70	(Kumar et al. 2020)
ZnO	H_2S	1	296	320/3592	(Hosseini et al. 2015)
CuO	H_2S	1	2.1	240/1341	(Li et al. 2017)
CO_3O_4	NH_3	100	146	2	(Wu et al. 2016a)
MoO_3	H_2	1000	90	14.1	(Yang et al. 2015a)
TiO_2	CH_4	60	6028	–	(Tshabalala et al. 2017)
ZnO	Ethanol	100	23	26/43	(Shankar and Rayappan 2017)
ZnO	Acetone	2	1.08	10/4	(Muthukrishnan et al. 2016)
SnO_2	Chloroform	300	–	~5.5 min/ ~7.5 min	(Abokifa et al. 2021)
Bi_2O_3	Toluene	50	15.66%	–	(Subin David et al. 2020)

$$R = \frac{R_g}{R_a} \text{ or } R = \frac{R_g - R_a}{R_a} \text{ or } R = \frac{R_g - R_a}{R_a} \times 100 \text{ for oxidizing gas; } R = \frac{R_a}{R_g} \text{ or } R = \frac{R_a - R_g}{R_g}$$

$$\text{or } R = \frac{R_a - R_g}{R_g} \times 100 \text{ for reducing gas}$$

The gas sensing performance was studied at room temperature and 250 °C and detected an excellent response at room temperature. The sensor showed a response of 296 at 1 ppm H_2S gas. However, it has also showed higher response/recovery times of 320/3592 s, respectively. Another NH_3 sensor based on 3D hierarchical porous p-type CO_3O_4 materials was also reported (Wu et al. 2016a). The material had the presence of many irregular structure defects with active sites, which actively participating in the reaction and thus improved the sensing response. The response of 146% at 100 ppm NH_3 was observed. The sensor was capable to detect gas concentration within 2 seconds with a low LOD of 0.5 ppm. Volatile organic compounds sensing was also done by using numerous metal oxides sensing. Mainly, n-type ZnO was widely utilized for the detection of volatile organic compounds (VOCs) (Shankar and Rayappan 2017). In brief, ZnO nanorods prepared by an electrospinning method were explored for ethanol sensing. At 200 ppm ethanol, the sensor showed a response around 23 s, with a lower rise and decay time of 26 s and 43 s, respectively.

5.2.3.2 Metal Decorated Metal Oxides for Gas Detection

The gas sensors by using numerous nanomaterials and their nanocomposites have been investigated and found very promising. However, there are certain limitations, such as limited sensitivity, higher working temperature, and selectivity issues. Therefore, there is a hunt for a new type of material and method to advance the gas sensor performance. In this scenario, the metal added nanomaterials could improve the sensitivity and selectivity due to the well-known spill-over effect. Table 5.4 shows gas sensor development based on a few metal-based nanocomposites for gas sensing.

Table 5.4 Summary of a few metal decorated metal oxides nanomaterials for gas sensing

Materials used	Gas analyte	Concen-tration (ppm)	Sensor response	Response/ Recovery time (s)	References
Au@SnO$_2$	NO$_2$	50	90	70	(Drmosh et al. 2018)
Au@ZnO/rGO	NO$_2$	1	67.38	248/170	(Cao et al. 2022)
Au@In$_2$O$_3$	CO	100	9	30/30	(Fu et al. 2017)
Pd@CuO	H$_2$S	100	1.196	-	(Kim et al. 2019)
Pt@VO$_x$	CH$_4$	500	18.2	~1000/~2000	(Liang et al. 2016)
Pd@TiO$_2$	H$_2$	8000	92.05	3.8/43.3	(Wei et al. 2017)
Ag@TiO$_2$	Ethanol	50	11.98	3/73	(Subha et al. 2017)

$$R = \frac{R_g}{R_a} \ \text{ or } \ R = \frac{R_g - R_a}{R_a} \ \text{ or } \ R = \frac{R_g - R_a}{R_a} \times 100 \text{ for oxidizing gas; } \ R = \frac{R_a}{R_g}$$

$$\text{or } \ R = \frac{R_a - R_g}{R_g} \ \text{ or } \ R = \frac{R_a - R_g}{R_g} \times 100 \text{ for reducing gas}$$

The metallic Au nanoparticles have been extensively used for finding various gases. For example, the Au nanoparticles decorated SnO$_2$ thin films were used for room temperature NO$_2$ sensing applications (Drmosh et al. 2018). The dynamic response curve was measured in the NO$_2$ concentration changed from 600 ppb, 2 ppm, 5 ppm, 8 ppm, and 11 ppm. The influence of temperature on the sensor performance and response time was also studied and observed that the sensor performance decreased with an increase in temperature at room temperature. The carbon monoxide gas sensing was also performed with Au decorated metal oxide nanomaterials. For example, an Au coated In_2O_3 rod-like structure was utilized for the CO sensing (Fu et al. 2017). The gas sensing response of different annealed samples was investigated and it was observed that the sample annealed at 300 °C, showed the best gas sensing results. The sensor showed a sensitivity of around 9 at 100 ppm CO gas and also displayed a fast-rising time of around 30 seconds. Additionally, sensor stability and humidity effect were also studied. Likewise, the magnetron sputtered Pt/VOx nanomaterial room temperature methane gas sensor was reported (Liang et al. 2016). The material was thermally annealed in O_2 atmosphere from 450 °C to 470 °C. The sensor annealed at 460 °C, showing a maximum sensing response of 18.2 to 500 ppm CH_4 gas, which was quite good at room temperature. The detection range was from 500 ppm to 2500 ppm and achieved a maximum response at 1500 ppm of CH_4 gas. The metal decorated sensory platform was also explored for volatile gas sensing. In detail, the Ag nanoparticles were also explored in gas sensing (Subha et al. 2017). The Ag-TiO_2 hybrid material was used to detect ethanol sensing. The TiO_2 nanorods were grown by a hydrothermal method and then the Ag nanoparticles deposited using a wet chemical deposition method. However, it was observed that after the addition of Ag, the sensor showed improved performance in terms of time characteristics rather than response. Also, the sensor detection limit was not much lower.

5.2.3.3 Metal-oxides with Metal Oxides Nanocomposites-based Gas Sensors

The n-n type metal oxides-based room temperature NOx sensor was reported. The SnO_2 contained oxygen vacancies on its surface and In_2O_3 improved the conductivity thus, both were responsible for the enhancement of the gas sensing performance (Xu et al. 2015). At room temperature, the response was not sufficient and observed up to 8.98 for 100 ppm NO_2 gas. Surprisingly, the response time was very fast, around 4.67 s with a detection limit of around 0.1 ppm. The In_2O_3/CuO type heterostructure (ICCN) was fabricated for NH_3 gas sensing (Zhou et al. 2018). The nanofibrous structure was synthesized by the electrospinning method. The TEM and HRTEM images are shown in Figures 5.2(a, b). It displayed the nanometer size mesoporous material along with lots of junctions like p-p homojunction and p-n type heterojunction on the material surface. The interaction of ICCN-5 with air, NH_3, and H_2S is displayed in Figure 5.2(c). The resistance of the sensor was measured from 100 ppm to 0.3 ppm NH_3 gas. The response time was just 2 seconds, while the response of the sensor was limited. The detection limit of around 0.3 ppm was reported.

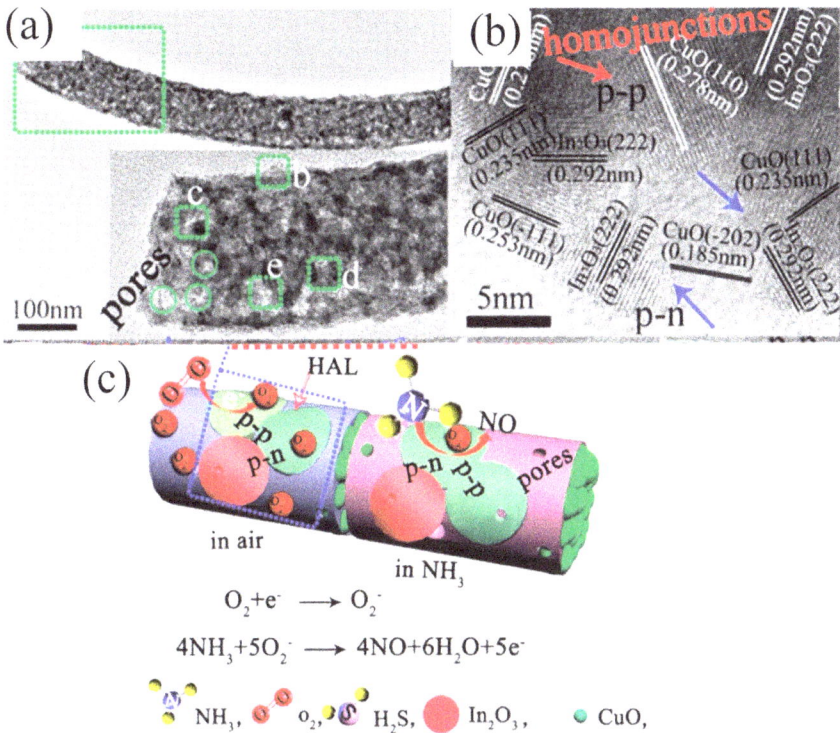

Figure 5.2 (a) TEM images of In_2O_3/CuO nanocomposite structure (ICCN-5); (b) HRTEM image, (c) Schematic of gas sensing of ICCN-5 and pristine CuO sensors to air and NH_3 or H_2S. [Reproduced with permission (Zhou et al. 2018). Copyright 2018, Elsevier].

Another new type of material based on Cr_2O_3-functionalized Nb_2O_5 nanostructured was tested for H_2 sensing (Park et al. 2016). The material film was prepared by using a hydrothermal method. The p-n nanocomposite sensor showed a response of 5.24 whereas, the Nb_2O_5 showed a response of 2.29 to 2 ppm hydrogen gas at RT. Nevertheless, the sensor displayed a longer recovery time of around 524.84 s. Typical ZnO/SnO_2 nanocomposite material was reported for ozone gas sensing at RT. The response was measured in ppb level of ozone gas (Silva et al. 2017). Regarding the change in % of SnO_2, the gas response was modified. The 50 wt% optimized SnO_2 in ZnO, achieved the response of 12 at 60 ppb with fast rising and decaying times of 13 s/90 s, respectively. The sensor showed a much lower detection limit of 20 ppb with a response value of 8. However, this sensor required a UV illumination source, which might be a limitation of the system. The literature on a few MOXs-MOXs based gas sensors is shown in Table 5.5.

Table 5.5 Summary of a few metal oxides-metal oxides composite-based gas sensors

Materials used	Gas analyte	Concentration (ppm)	Sensor response	Response/ Recovery time (s)	References
N-CoS$_2$/CO$_3$O$_4$	NO$_2$	100	6230%	1.3/17.98	(Bai et al. 2022)
In$_2$O$_3$/SnO$_2$	NO$_x$	100	8.98	4.67	(Xu et al. 2015)
Cu$_2$O/CuO	NO$_2$	0.5	825%	66/1020	(Wang et al. 2021a)
In$_2$O$_3$/CuO	NH$_3$	100	1.9	2	(Zhou et al. 2018)
WO$_3$/rGO	H$_2$S	0.5	32.7	340/180	(Peng et al. 2020)
Cu$_2$O/CO$_3$O$_4$	H$_2$S	20	~2600	~100/~100	(Cui et al. 2017)
Cr$_2$O$_3$/Nb$_2$O$_5$	H$_2$	200	5.24	40	(Park et al. 2016)
ZnO/SnO$_2$	Ozone	0.06	12	13/90	(Silva et al. 2017)
α-Fe$_2$O$_3$/ZnO	Ethanol	700	706.8	–	(Zhu et al. 2016)
VO$_2$/ZnO	Acetone	100	4.51	8/18	(Liang et al. 2018)

$$R = \frac{R_g}{R_a} \text{ or } R = \frac{R_g - R_a}{R_a} \text{ or } R = \frac{R_g - R_a}{R_a} \times 100 \text{ for oxidizing gas; } R = \frac{R_a}{R_g} \text{ or } R = \frac{R_a - R_g}{R_g}$$

$$\text{or } R = \frac{R_a - R_g}{R_g} \times 100 \text{ for reducing gas}$$

5.2.3.4 Metal Oxide Nanostructures and Carbon Nanomaterials

At present, MOXs-based carbon based heterostructures are very promising for gas sensing. Various types of carbon structures, such as rGO, GO, CNTs, and graphene were explored for sensing reducing and oxidizing gases along with numerous VOCs. The summary of a few gas sensors based on metal oxides-carbon materials composites is tabulated in Table 5.6.

A highly sensitive SnO$_2$ nanoparticles-rGO hybrid was utilized for NO$_2$ sensing (Wang et al. 2018). The SnO$_2$ nanoparticles were prepared by using a hydrothermal method. The XPS study confirmed the SnO$_2$ NPS contains a lot of oxygen vacancies, which actively participated in the reaction kinetics and improved the gas response. At 1 ppm NO$_2$, the sensor showed a high response of 3.8, with a rising and decaying time of 14 s. But the device decline time was little longer of around 190 s. With respect to different synthesis conditions, the response and time characteristics were modulated. This low limit detected sensor could be useful in the gas sensing field. Similarly, another carbon structure like CNTs was also used for gas sensing. In detail, the hydrothermally synthesized (VO$_2$–CNT) nanocomposite material was prepared and tested to 20–45 ppm NH$_3$ (Evans et al. 2018). The device performance was also verified at different humidity environments. The sensor response was not significant and also showed longer response and recovery times, suggesting its unsuitability for gas sensing applications. A room temperature Pd-decorated SnO_2/rGO nanocomposite was utilized for CO detection (Shojaee et al. 2018). The spill-over effect of Pd on SnO_2 nanoparticles enhanced the sensing performance. Still, the response was also not up to the mark. The sensor showed a response of 9.5 at 1600 ppm of CO gas. The response and recovery times were also a little longer 2 minutes each. However, the simple synthesis method could be used for fabrication of low-cost sensor platform. The electrochemical deposition method was successfully carried out for synthesizing SnO_2/graphene structure and used to detect formaldehyde gas (Bo et al. 2018). The sensor has shown a linear response in 0 to 210 ppm of HCHO gas concentration and achieved a remarkable low LOD of 0.02 ppm. The device stability of over 12 days and the repeatability test confirmed its good stability. But the sensor response was a little affected when it was tested at different temperatures.

Table 5.6 A summary of a few gas sensors based on metal oxides-carbon materials composites

Materials used	Gas analyte	Concentration (ppm)	Sensor response	Response/ Recovery time (s)	References
rGO/α-Fe$_2$O$_3$	NO$_2$	5	8.2	2.1 min/ 40 min	(Zhang et al. 2018)
SnO$_2$/rGO	NO$_2$	1	3.8	14/190	(Wang et al. 2018)
VO$_2$/CNT	NH$_3$	45	0.04	290/1800	(Evans et al. 2018)
Pd@SnO$_2$/rGO	CO	1600	9.5	2 min/ 2 min	(Shojaee et al. 2018)
ZnO/3D-RGO	CO	1000	27.5%	14/15	(Ha et al. 2018)
SnO$_2$/rGO	H$_2$S	50	33	2/292	(Song et al. 2016)
ZnO/graphene	H$_2$	100	28.08	30/~150	(Kathiravan et al. 2017)
SnO$_2$/graphene	HCOH	5	4.6	46/95	(Bo et al. 2018)

$$R = \frac{R_g}{R_a} \text{ or } R = \frac{R_g - R_a}{R_a} \text{ or } R = \frac{R_g - R_a}{R_a} \times 100 \text{ for oxidizing gas; } R = \frac{R_a}{R_g} \text{ or}$$

$$R = \frac{R_a - R_g}{R_g} \text{ or } R = \frac{R_a - R_g}{R_g} \times 100 \text{ for reducing gas}$$

5.2.3.5 Ultraviolet Light-based Gas Sensing

Different strategies have been used to improve gas sensing performance. Among that, UV light illumination has been widely used because it significantly enhances the sensor response, selectivity and response, recovery time etc. The development of few UV-illuminated gas sensors is shown in Table 5.7. The details of certain gas sensors in this area are as follows.

A NO_2 sensor was reported based on g-C_3N_4/GaN nanocomposite synthesized by vacuum filtration method (Reddeppa et al. 2021). At 1.35 mw/cm^2 lighting, the sensor showed a response of around 7.8% at 5 ppm NO_2 gas. Only the g-C_3N_4 sample did not exhibit a significant response. The sensor was able to detect 500 ppb of NO_2 gas. Surprisingly, the sensor response time was shorter than the recovery time. The CuPc-loaded ZnO nanorod structure synthesized by the microwave assisted hydrothermal route was utilized for ammonia sensing (Huang et al. 2021). When red light shined on the sensing surface, the sensor was activated and showed excellent response, dynamic response characteristics, and excessive selectivity to ammonia gas. Very low concentration of ammonia gas of 0.8 ppm could be detected by this sensor. The sensor presented a good response and recovery time of 20 s and 10 s, with a response of 15.8 at 100 ppm of ammonia. Despite the good performance, one thing is that the ammonia concentration was a little high at around 100 ppm. There are a few other sensors also reported at lower concentration of ammonia. The GaN nanorods combined with the rGO nanosheets were examined for H_2S sensing (Reddeppa et al. 2018). Under light illumination of 365 nm, the sensor response was significantly improved and achieved around 15.20%. The sensor performance was also investigated for GaN nanorods. It was verified that the nanocomposite was a perfect material for H_2S sensing. The humidity effect on the sensor performance was also studied. Though, the concentration of H_2S was high and the sensor recovery time was much higher than expected and thus, needs to be improved. The nanospheres were also widely used in the gas sensing field. In this domain, the hollow TiO_2/SnO_2 nanospheres prepared by the hydrothermal method were explored for HCHO sensing (Hsu et al. 2018). A high response of 20 was detected at 10 ppm HCHO concentration. The well-formation of hetero junction at the material interface was responsible to show this high sensing performance. The sensor displayed a fast temporal characteristic that could be the advantage of this device. The detection limit was also in the range of ppb and thus, signifying its use for gas sensing applications. The hydrothermally synthesized Na-doped ZnO nanowires were studied for ethanol sensing applications (Hsu et al. 2018). This p-type material was able to detect ethanol in the concentration range of 5 to 60 ppm with a corresponding response of –29 to –40%. However, this sensor recovery time was also a little longer of the order of >10 min. UV-activated ZnO nanosheets based methane gas sensor at room temperature was also reported (Wang et al. 2021b).

Table 5.7 Summary of a few gas sensors based on UV-light illumination

Materials used	Gas analyte	Concentration (ppm)	Sensor response	Response/ Recovery time (s)	References
ZnO	NO_2	20	411	221/118	(Meng et al. 2018)
g-C_3N_4/GaN	NO_2	5	7.8%	59/1146	(Reddeppa et al. 2021)
Ti_3C_2Tx Mxene/ZnO	NO_2	200 ppb	346%	18/26	(Wang et al. 2022)
CuPc@ZnO	NH_3	100	15.8	20/10	(Huang et al. 2021)
rGO/GaN	H_2S	100	15.20%	~200/~1100	(Reddeppa et al. 2018)
rGO/ZnO/Pd	CH_4	10000	63.4%	74/78	(Xia et al. 2021)
rGO/SnO_2	SO_2	5	11%	4.3 min/ 2.5 min	(Li et al. 2019)
rGO/ZnO	H_2	500	96%	8/612	(Drmosh et al. 2019)
TiO_2/SnO_2	HCHO	10	20	20/56	(Zhang et al. 2020a)
p-type Na-doped ZnO	Ethanol	6	−40	72/>10 min	(Hsu et al. 2018)
Graphene	Acetone	1	1.87%	4 min	(Yang et al. 2017)

$$R = \frac{R_g}{R_a} \text{ or } R = \frac{R_g - R_a}{R_a} \text{ or } R = \frac{R_g - R_a}{R_a} \times 100 \text{ for oxidizing gas; } R = \frac{R_a}{R_g} \text{ or } R = \frac{R_a - R_g}{R_g}$$
$$\text{or } R = \frac{R_a - R_g}{R_g} \times 100 \text{ for reducing gas}$$

5.2.3.6 Gas Sensing at a High Working Temperature

The combination of two different metal oxides leads to high gas adsorption, large surface defects, more heterojunction formation, an increase in gas conductivity, decrease in working temperature, improvement in response and recovery time, and enhancement in cross-selectivity, which are very important for practical gas-sensing applications. The development of gas sensors at high working temperature is shown in Table 5.8.

The ZnO-rGO used NO_2 chemiresistive gas sensor was also reported (Cao et al. 2020). The ZnO nanospheres were synthesized by using a solvothermal method. The ZnO nanospheres were uniformly dispersed onto the surface of the rGO nanosheet, and therefore, increased good surface contact and surface area. The heterojunction present on the surface initiates the good flow of charge carriers, and therefore, amplified

Table 5.8 Summary of a gas sensors at high working temperature

Materials used	Gas analyte	T (°C)	Concen- tration (ppm)	Sensor response	Response/ Recovery time (s)	References
WO$_3$/rGO	NO$_2$	100	5	9.67%	180/432	(Jang et al. 2019)
ZnO/rGO–1	NO$_2$	110	2.5	33.11	182/234	(Cao et al. 2020)
SnO$_2$/WO$_3$	NH$_3$	300	250	7.1	12/58	(Toan et al. 2017)
Ag$_3$PO$_4$	NH$_3$	50	100	52%	276/1338	(Yan et al. 2019)
α-Fe$_2$O$_3$/ graphene	NH$_3$	250	10	13.5%	152/10.8 min	(Haridas et al. 2020)
CuO/8%g–C$_3$N$_4$	CO	150	4000	20%	100/60	(Nasresfahani et al. 2021)
ZnO/rGO	CO	200	200	85.2%	7/9	(Ha et al. 2018)
SnO$_2$/CuO	CH$_4$	150	500	6	< 1 min/ < 1 min	(Das and Panda 2019)
CeO$_2$/SnO$_2$	H$_2$	300	60	1323	17/24	(Liang et al. 2022)
SnO$_2$	TEA	270	100	49.5	14/12	(Zou et al. 2017)
ZnFe$_2$O$_4$	Toluene	250	100	79		(Liu et al. 2022)
NiO/NiCr$_2$O$_4$	Xylene	225	100	66.2	1217/591	(Gao et al. 2019)
SnO$_2$/Zn$_2$SnO$_4$	Ethanol	250	100	300	2/114	(Yang et al. 2019)
ZnO/SnO$_2$	Ethanol	400	25	50	–	(Tharsika et al. 2019)
rGO–/ZnO/ SnO$_2$	Acetone	150	10	91%	10/100	(Sen and Kundu 2021)
Au–O–SnO$_2$–5	DMMP	320	680 ppb	1.67	26/32	(Yang et al. 2022)

$$R = \frac{R_g}{R_a} \text{ or } R = \frac{R_g - R_a}{R_a} \text{ or } R = \frac{R_g - R_a}{R_a} \times 100 \text{ for oxidizing gas; } R = \frac{R_a}{R_g} \text{ or } R = \frac{R_a - R_g}{R_g}$$

$$\text{or } R = \frac{R_a - R_g}{R_g} \times 100 \text{ for reducing gas}$$

the sensing response. A remarkable low LOD of 5 ppb was achieved. However, the sensor working temperature and LOD is a little higher and needs to be decreased further. A chemiresistive-based NH$_3$ sensor was also studied. The strong interaction between the π-network with α-Fe$_2$O$_3$ was responsible to enhance the gas sensing response (Haridas et al. 2020). The sensor showed excellent response, repeatability, and selectivity to NH$_3$ gas. Although, the reported sensor showed a good sensitivity of 13.5%, the recovery time was around 10.8 minutes, which limits its use for practical applications. A highly sensitive CO sensor was prepared hydrothermally by utilizing ZnO/rGO nanocomposite materials (Ha et al. 2018). The response and decay periods of less than ten seconds were reported. The gas sensing performance was also tested at room temperature and observed the response around

27.5% to 1000 ppm CO gas. The sensor exhibited a good selectivity to CO gas and stability over weeks. However, the CO gas concentration was significantly higher than 1000 ppm and the response needs to be improved further. A trace level of methane was detected by SnO_2-CuO nanocomposite synthesized by using a co-precipitation chemical route followed by annealing in air (Das and Panda 2019). The sensor response was measured at a much lower temperature of 150 °C. The pristine annealed SnO_2 sensor was also tested on CH_4 but showed a poor sensing performance. The nanocomposite showed a good response at 500 ppm of CH_4 gas. Interestingly, the response and recovery time of less than a minute was reported. However, the sensor cross-selectivity to H_2 alone was reported. The hollow urchin-like core-shell $ZnFe_2O_4$ spheres prepared using a hydrothermal treatment were utilized to detect toluene (Liu et al. 2022). The urchin-like core-shell structure exhibited a high surface to volume ratio and oxygen vacancies and chemisorbed oxygen species and thus, exposed a high gas sensing performance with an LOD of 0.2 ppm being reported. Though, under a humidity concentration of 10–98% RH, the response decreased by 30%. In another work of VOCs sensing, the sensitivity of ethanol sensor was improved by combining porous structure SnO_2/Zn_2SnO_4 nanocomposite (Yang et al. 2019). The proposed material was synthesized by hydrothermal method and explored for ethanol detection from 0.5 to 100 ppm. The heterojunction and porous nature of the material greatly enhanced the sensing response of 300 at 250 °C optimum temperature. The sensor showed a significant response of 1.4 at 0.5 ppm. Still, the high working temperature of the device could be the limitation of the work.

Various strategies have been used to improve the gas sensing response. The MOX-MOX based composites have been explored for gas sensing very effectively. However, the working temperature of the device is on the higher side. Further strategies have been used to improve the response by using MOX-polymeric based nanocomposites for gas detection. These types of sensors can work at room temperature and could show high response and good time-characteristics. However, due to its polymeric nature, it might be prone to water vapor. Therefore, additional approaches are required to increase the response of the sensor.

5.2.3.7   Conducting Polymers for Gas and VOCs Sensors

At present, organic conducting polymers are promising candidates for the detection of various gases and vapors. The unique advantages such as room temperature operation, simple synthesis method, and ease in sensing geometry preparation have helped researchers to choose certain polymers for gas sensing applications. Table 5.9 shows a summary of polymer-based gas sensors.

Table 5.9 Gas sensors based on polymers

Materials used	Gas analyte	Detection range	Concentration (ppm)	Sensor response	Response/ Recovery time (s)	References
Polypyrrole/ nitrogen-doped multiwall carbon nanotube	NO_2	0.25–9 ppm	5	24.82%	65/668	(Liu et al. 2019a)
PEDOT	NO_2	10–100 ppm	100	22% change of resistance	–	(Dunst et al. 2017)
SnO_2/ polypyrrole	NH_3	1–10.7 ppm	10.7	75%	259/468 @5 ppm	(Li et al. 2016)
ZnO-en-PPy	NH_3	1–100 ppm	100	0.4947 $k\Omega$/ppm	45/55	(Singh et al. 2021)
Polyaniline/ $SrGe_4O_9$	NH_3	0.2–10 ppm	0.2	16%	24/–@800 ppb	(Zhang et al. 2020b)
Polythiophene/ chitosan	NH_3	@100 ppm	100	~0.37	–	(Amruth et al. 2022)
MWCNTs/ PANI	NH_3	33–100 ppm	50	117%	47	(Ma et al. 2021)
PPy-rGO	DMMP	5–100 ppm	100	12.9%	43/75	(Yang et al. 2021)
PANI–PEO	H_2S	1–10 ppm	10	25%	120/250	(Mousavi et al. 2016)

$$R = \frac{R_g}{R_a} \text{ or } R = \frac{R_g - R_a}{R_a} \text{ or } R = \frac{R_g - R_a}{R_a} \times 100 \text{ for oxidizing gas; } R = \frac{R_a}{R_g} \text{ or } R = \frac{R_a - R_g}{R_g}$$

$$\text{or } R = \frac{R_a - R_g}{R_g} \times 100 \text{ for reducing gas}$$

In situ, a self-assembled nitrogen dioxide sensor using a polypyrrole/ nitrogen doped multiwall carbon nanotube was fabricated (Liu et al. 2019a). The material annealed at 350 °C and its room temperature response was observed at around 24.82% to 5 ppm NO_2 gas. The sensor detection range was 0.25–9 ppm. The high recovery time of several minutes might be the limitation of this work. Another polymer polypyrrole was used to detect the gas analyte. The SnO_2/PPy nanocomposite was synthesized using vapor phase polymerization of pyrrole on the substrate having SnO_2 nanosheets (Li et al. 2016). The p-n heterojunction between the SnO_2 and polypyrrole increases the surface contact and surface area, which further raised the sensing performance. The sensor showed a sensing response of 75% to a low concentration of 10.7 ppm and an LOD of ~257 ppb of NH_3 gas was observed. A flexible H_2S sensor using polyaniline–polyethylene oxide (PAni–PEO) nanofibers doped by camphorsulfonic acid (HCSA) was reported (Mousavi et al. 2016). The sensing geometry was fabricated on a flexible paper and polyamide substrate to which the H_2S gas was exposed. The sensor stability, humidity effect and selectivity dependence were investigated. However, in the cross-selectivity study, only two gases NO_2 and acetone were considered.

5.2.3.8   Metal-organic Frameworks-based Gas Sensors

A room temperature NO_2 gas sensor based on the Fe atoms in Fe_3-PCN-250 MOFs was investigated (Khan et al. 2022). In this structure, the different transition metals like Co, Mn, and Zn were partially replaced via a hydrothermal route, which changed its structural and electrical properties. This study reported that the Fe_2Mn PCN-250 exhibited a higher gas sensing response than the pure structure. Still, the sensor time characteristics were much higher (1325 s/5184 s), which limits its use for practical applications. The high density of the metal cluster was used to improve the gas sensing response. In detail, the Co-incorporated MOF was used to detect H_2 (Nguyen et al. 2020). The highly dense Co ions (+2 and +3 states) situated in metallic groups are responsible for the Co-MOF-74 response enhancement to H_2. The response performance was also compared with Ni-MOF-74 and Mg-MOF-74. At 50 ppm H_2, the sensor exhibited a response of 101.4%. The cross-selectivity study to other interfering gases was also investigated. The different ratios of Ni and Zn were synthesized by calcinating the metal-organic framework (ZnOF-x). A MOF-derived ZnO-NiO nanocomposites for acetone sensing were studied (Sun et al. 2022). It was detected that the ZnO-x showed a lower sensitivity but better moisture-resistive response to acetone by increasing the amount of Ni. Under humidity of 95% RH, the ZnO-5 exhibited a higher gas sensing response of 1.31 to 1 ppm NO_2 gas. However, the working temperature of the device was also higher than 175 °C. The ZIF-67 was also tested for the detection of p-xylene (Jo et al. 2018). The hollow CO_3O_4 nanocages having 4 dissimilar dimensions (~0.3, 1.0, 2.0, and 4.0 μm) containing nanosheets were synthesized. Hollow hierarchical CO_3O_4 nanocages with a size of ~1.0 μm exhibited a massive resistance ratio of 78.6 to p-xylene at 225 °C. The work did not report the stability study of the material.

Following Table 5.10 shows some development based on metal-organic frameworks.

Table 5.10 Summary of a few gas sensors based on metal-organic frameworks

Materials used	Gas analyte	T (°C)	Concentration (ppm)	Sensor response	Response/ Recovery time (s)	References
Fe$_2$Mn PCN-250	NO$_2$	RT	0.5	~2.7%	1325/5184	(Khan et al. 2022)
Fe$_2$O$_3$–Fe–PBA	H$_2$S	200	5	30.8	–	(Guo et al. 2019)
CuO/In$_2$O$_3$– In-MOF CPP-3(In)	H$_2$S	70	5	229.3	10	(Li et al. 2020b)
Co-MOF-I	H$_2$	200	10	~17%	~300/~300	(Nguyen et al. 2020)
Co/Zn doped carbon nanotube-ZIF-67	SO$_2$	RT	30	28.9	78/900	(Li et al. 2018)
MOFs with Zinc and Nickel	NH$_3$	RT	50	291	8	(Mohan Reddy et al. 2022)
MOF-derived ZnO-NiO	Acetone	175	1	1.31	–	(Sun et al. 2022)
MOFs with Zinc and Nickel	Ethanol	RT	50	~5	28	(Mohan Reddy et al. 2022)
MOF-derived porous CO$_3$O$_4$	*n*-butanol	140	100	53.78	99/50	(Cheng et al. 2021b)
CO$_3$O$_4$-ZIF-67	p-Xylene	225	5	resistance ratio 78.6	63/86	(Jo et al. 2018)
Cu-doped Fe$_2$O$_3$–MIL-88(Fe)	TEA	240	100	1.7	2/7	(Gao et al. 2021)
CuO/ZnO– Zn–MOF HPU-15	Methanol	260	500	19	20/300	(Li et al. 2020a)

$R = \dfrac{R_g}{R_a}$ or $R = \dfrac{R_g - R_a}{R_a - R_g}$ or $R = \dfrac{R_g - R_a}{R_a} \times 100$ for oxidizing gas; $R = \dfrac{R_a}{R_g}$ or $R = \dfrac{R_a - R_g}{R_g}$ or $R = \dfrac{R_a - R_g}{R_g} \times 100$ for reducing gas

5.3 FIELD-EFFECT TRANSISTOR (FET)-BASED GAS SENSORS

At present, FETs-based technology has attracted considerable attention in the gas sensing field. The numerous features like low power consumption, fine tunability with advanced technology, and good response values are the key features of this technology. Table 5.11 shows a summary of FETs-based gas sensors.

Table 5.11 Development of a few field effect transistors-based gas sensors

Materials used	Gas analyte	T (°C)	Detection range	Sensor response	Response/ Recovery time (s)	References
ZnO	NO_2	180	1–20 ppm	176% @20 ppm	90/700	(Hong et al. 2016)
Gr/MoS$_2$	NO_2	RT	@1 ppm	10^3	–	(Tabata et al. 2018)
Indium-gallium-ZnO	NO_2	150	@500 ppb	~65%	–	(Shin et al. 2022)
CNT$_S$	H_2	RT	0.02%–0.1%	–	–	(Jeon et al. 2017)
PEI-CNT$_S$	CO_2	RT	0.005–1%	~1.4%	10/20	(Cho et al. 2018)
S6DNA-SWCNTs	4-ethyl phenol	RT	@0.13%	–	–	(Wang et al. 2020)
Pd-POFs-SiNW$_S$	methanol	RT	1200–6400 ppm	–	–	(Cao et al. 2018)
ZnO	ethanol	RT	@100 ppm	3	–	(Ponhan et al. 2017)
ZnO	H_2S	180	@20 ppm	58%	–	(Hong et al. 2016)
Pd/GO	acetone	RT	@0.4 ppm	15%	37/91	(Bhardwaj and Hazra 2022)
Ni$_3$BTC$_2$–OH–SWCNTs	SO_2	RT	4–20 ppm	–	4.59/11.04	(Ingle et al. 2020)

$$R = \frac{I_{DS,\text{gas}} - I_{DS,\text{air}}}{I_{DS,\text{air}}} \ \text{ or } \ R = \frac{I_{DS,\text{gas}}}{I_{DS,\text{air}}} - 1 \text{ or } R = \frac{I_{DS,\text{gas}} - I_{DS,\text{air}}}{I_{DS,\text{air}}} \times 100$$

The NO_2 gas detection has been implemented by using various FET-based configurations. In detail, the ZnO-based NO_2 gas sensor was prepared by depositing a sensitive ZnO layer by atomic layer

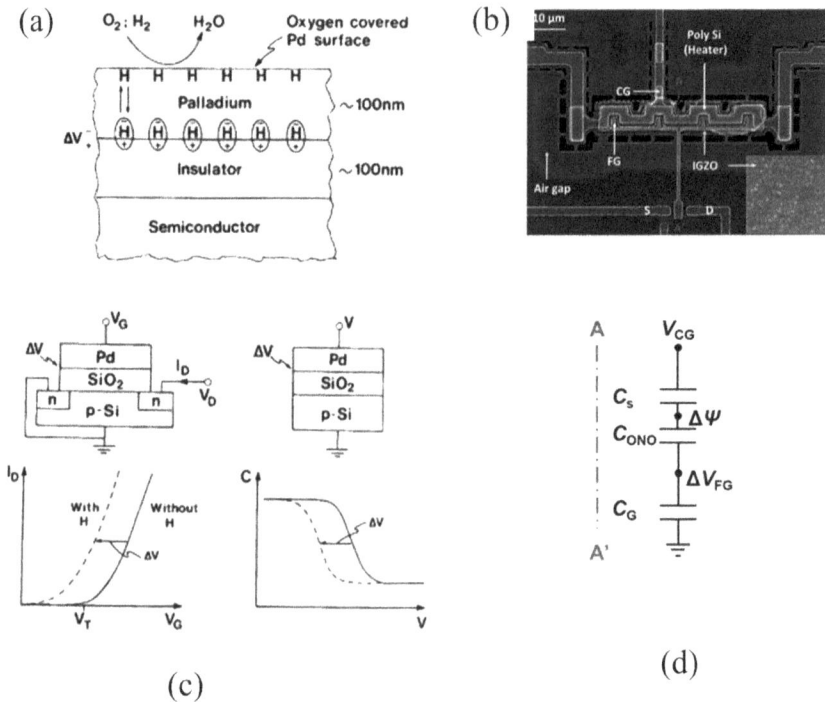

Figure 5.3 (a, c) Schematic of Pd-metal oxides-based hydrogen sensitive field-effect device. The sensor characteristics in terms of a shift in the voltage upon the interaction of H_2 gas at the metal–oxide interface. [Reproduced with permission (Lundström et al. 2007). Copyright 2007, Elsevier]; (b) SEMof the FET-type gas sensor with a micro-heater, and (d) Basic capacitive model of the sensor. [Reproduced with permission (Shin et al. 2022). Copyright 2022, Elsevier].

deposition technique (Hong et al. 2016). As NO_2 being oxidizing by nature, it significantly alters the charge carrier concentration of n-type ZnO and thus, the response of 176% at 20 ppm of NO_2 was obtained. Since, most of the MOX are effectively working at a higher temperature, therefore, in this report, the material working temperature was set at 180 °C. Additionally, the response time was in seconds but the recovery time was in minutes. Other semiconductor and metal materials were used for H_2 sensing (Lundström et al. 2007). The metal palladium and silicon dioxide and the possible interaction between hydrogen gas with metal-catalytic interface are shown in Figure 5.3(a, d). The graph shows the voltage drift with and without hydrogen. Another MOX was also explored for monitoring NO_2. The FET structure was coated with an indium-gallium-zinc oxide (IGZO) material as a sensing layer

(Shin et al. 2022). The top SEM image and the equivalent capacitive model of the FET-type gas sensor with the embedded micro-heater are shown in Figure 5.3(b, d). The buried channel showed ~10 times higher SNR than the surface channel because of the small 1/f noise. The SNR can be realized by correctly selecting the operation region. This device has shown a very low LOD of 44.2 ppt.

At present, CNTs are attracting a lot of interest in the gas sensing field because of their great conductivity and high surface area (Jeon et al. 2017). The property of CNTs was used for H_2 sensing. Here, the CNTs were used as the sensing layer and the Pd used as a source, drain, and gate. By using this structure, the H_2 concentration from 0.02% to 0.1% was monitored at room temperature. The sensor response was only studied with H_2 and NO_2 gas. The surface morphology of the CNTs was modified by the functionalization technique (Cho et al. 2018). In this, the CNTs-based sensing device was used to monitor the CO_2 gas in the concentration of 0.005–1%. The CNTs surface was functionalized with polyethylenimine (PEI) with rich amine groups to selectively bind the gas molecules. A response of around 1.4% was observed for CO_2 gas. Still, the response needs to improve further. The loading of metal nanoparticles into certain matrix has also been considered. In short, the metal nanoparticles such as Au, Pd, and Pt were mixed into GO and utilized for vapor sensing (Bhardwaj and Hazra 2022). These metallic nanoparticles were coated by using a spray coating method. The metals helped in decreasing the sensors working temperature. Very low acetone gas sensing was studied at 50 °C. Among these gas sensor devices, the Pd/GO showed the best gas sensing results. The sensor was capable of detecting detect 400 ppb of acetone concentration. The key feature of this work includes low working temperature, low response and recovery time in tens of seconds, and low concentration detection.

5.4 OPTICAL FIBER SENSORS

At present, monitoring gases and vapors are becoming an urgent need for human and environmental safety. In this regard, the optical fiber has been playing an important role due to its capability of on site parameter monitoring, easy to handle due to less requirement of the optical parts, can carry a long range of wavelengths, and miniaturized size, etc. The typical fiber optics-based gas sensor geometries were explored as shown in Figure 5.4. The development of a few optical fiber gas sensors is tabulated in Table 5.12.

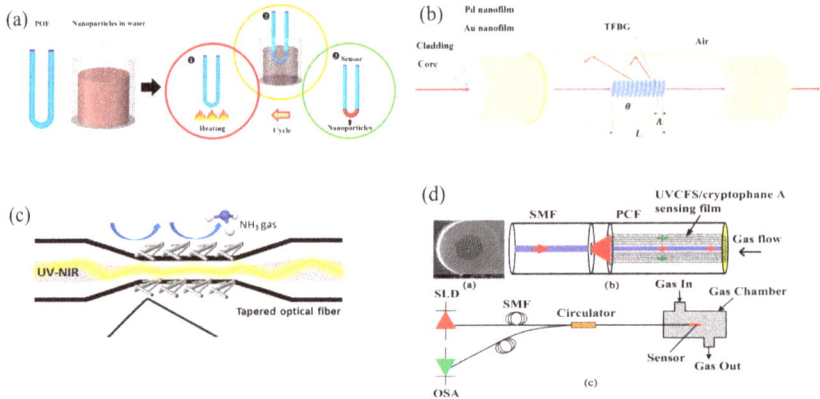

Figure 5.4 Different types of optical fiber geometries used for gas sensing. (a) U-shape optical fiber coated with copper and iron oxide nanoparticles for H_2S sensing, [Reproduced with permission (Lopez et al. 2021). Copyright 2021, Elsevier]; (b) Pd/Au nanofilms coated tilted fiber Bragg grating for hydrogen sensing, [Reproduced with permission (Zhang et al. 2022d). Copyright 2022, Elsevier]; (c) The metal oxide h-MoO$_3$ coated on tapered section of optical fiber for NH$_3$ sensing, [Reproduced with permission (Chua et al. 2021). Copyright 2021, Elsevier]; and (d) An ultraviolet curable fluoro-siloxane nano-film incorporating cryptophane A coated photonic crystal fiber for CH_4 sensing, [Reproduced with permission (Yang et al. 2016). Copyright 2016, Elsevier].

Table 5.12 Summary of a few optical fiber integrated with numerous nanomaterials for various gases and vapors detection

Materials used	Gas analyte	Detection range	Concen-tration	Sensor response	Response/ Recovery time (s)	References
MoS$_2$/citric acid coated LPG fiber	H$_2$S	0–70 ppm	0.5 ppm	10.52 pm/ ppm	89/97	(Qin et al. 2018)
ZnO/PMMA matrix	H$_2$S	1–5	–	1950 pm/ ppm	–/27	(Kitture et al. 2017)
Cu$_{2-x}$O	H$_2$S	10–200 ppm	10 ppm	15.11% @50 ppm	1 min	(Keley et al. 2021)
Fe$_3$O$_4$ coated POF U-shape	H$_2$S	50–200 ppm	35 ppm	0.6%@200 0.6%	~376/~300	(Lopez et al. 2021)
Pd/WO$_3$ coated on PCF	H$_2$	0–10,000 µl/l	440.367 µl/l	0.109 pm/ (µl/l)	<33 min	(Zhang et al. 2016)

(Contd.)

Materials used	Gas analyte	Detection range	Concentration	Sensor response	Response/ Recovery time (s)	References
Pd/SiO$_2$ coated on uncladded MMF	H$_2$	100–2000 ppm	100 ppm	–C	~5 min/ 30 min	(Sun et al. 2017)
Pt@WO$_3$ coated fiber	H$_2$	0.5–5%	495 ppm	–1.063 V/%	14.48 min	(Wei et al. 2022)
PDMS/WO$_3$ coated tilted FBGs	H$_2$	0–1.53%	–	0.596 dB/%	93/107	(Zhang et al. 2022b)
Pd/Au coated FBGs	H$_2$	0–1.02%	–	1.597 dB%	37/49	(Zhang et al. 2022d)
Polycarbonate/Cryptophane A coated on LPG	CH$_4$	0–3.5%	0.2%	2.5 nm/%	–	(Yang et al. 2015b)
Graphene-SnO$_2$ coated on uncladded SMF	CH$_4$	0–55%	–	200 a.u/%	–	(Zhang et al. 2017b)
PCF coated film	CH$_4$	0–2%	–	0.794 nm/%	–	(Liu et al. 2020)
ZnO coated on uncladded MMF	NH$_3$	0–500 ppm	–	24 × 10^{-3}/ kPa	11 min/ 8 min	(Devendiran and Sastikumar 2017)
Ag@ PVP/PVA coated on uncladded POF	NH$_3$	0–500 ppm	–	0.88 counts/ ppm	~90 min/ ~90 min	(Rithesh Raj et al. 2015)
Graphene and Fe$_3$O$_4$/ Graphene coated at tip of SMF fiber	NH$_3$	1.5–150 ppm	10/7 ppb	25/36 pm/ ppm	–/6–8	(Pawar et al. 2018)
Polyaniline coated silica optical fiber	NH$_3$	0.125–1%	–C	2.2 to 1%	2.82 min/ 11.52 min	(Ibrahim et al. 2022)
h-MoO$_3$ coated MMF	NH$_3$	100–5000 ppm	500 ppm	89.13% @10,000 ppm	210/241	(Chua et al. 2021)

(Contd.)

Table 5.12 Summary of a few optical fiber integrated with numerous nanomaterials for various gases and vaporsdetection (*Contd.*)

Materials used	Gas analyte	Detection range	Concen-tration	Sensor response	Response/ Recovery time (s)	References
Fe_2O_3 coated tapered microfiber	NH_3	0–10,476 ppm	–	1.30 pm/ ppm	–	(Fu et al. 2020)
SnO_2 coated microfiber	NH_3	0–10,476 ppm	–	0.58 pm/ ppm	–	(Fu et al. 2022)
ZnO coated LPG created in SMF	CO_2	3–15%	–	0.0513 dB/%	~6 min/ ~20 min	(Wu et al. 2016b)
rGO coated etched FBG in SMF	NO_2	0.5–3 ppm	0.5 ppm	34.3 pm @3ppm	12.3 min/3.8 min	(Sridevi. et al. 2016)
GO coated PM microfiber	Ethanol	0–80 ppm	1.6 ppm	0.138 nm/ ppm	6.1 min	(Zhang et al. 2017a)
TiO_2/GO coated tip of the MMF	Ethylene	0–909 ppm	–	0.92 pm/ ppm	–	(Tabassum et al. 2017)
CeO_2 doped nanocrystal-line SnO_2 clad modi-fied PMMA fiber	Benzene	0–1000 ppm	–	-79 counts/100 ppm	51 and 37 min	(Rengana-than et al. 2021)
	Metha-nol		–	23 counts/ 100 ppm		
	Chloro-form		–	13 counts/ 100 ppm		
ZnO coated at distal end of SMF	Gaso-line blend mix-tures	0–100%	–	12.1 nm	5/9	(Pawar et al. 2017)
5%K-Fe_2O_3 coated PMMA MMF	Ethanol	0–500 ppm	–	7.5×10^{-3}/ kPa	56/73	(Kalai Priya et al. 2020)

$R = \dfrac{l_g - l_0}{l_0}$ or $R = \dfrac{l_g - l_0}{l_0} \times 100$ for wavelength shift; $R = \dfrac{P_g - P_0}{P_0}$ or $R = \dfrac{P_g - P_0}{P_0} \times 100$ for power shift

5.4.1 H₂S Sensor

H_2S is extremely toxic and mainly produced by gas plants, sewage plants, well water, and the oil industry. The exposure limit of H_2S is 20 ppm.

The exposure of high ppm of H_2S (>100 ppm) can lead to death. Therefore, the monitoring of H_2S is very essential for human safety and the environment.

It is observed that numerous materials such as MOXs, 2D materials, and different polymers are widely employed for H_2S detection. Due to the strong reducing nature of H_2S, it easily interacts with O_2 ion species present on the material's surface and this ensures easy transfer of electrons. This leads to alteration in the dielectric constant of the material, and therefore, changes in the refractive index of the material. Thus, it is a very easy way to monitor the minute concentration of the H_2S gas. Fiber based LPG type was used for the detection of H_2S (Qin et al. 2018). The Molybdenum sulfide/citric acid nanocomposite material was coated on the top of the LPG and its wavelength shift recorded with H_2S concentration of 0–70 ppm. The blue shift was observed in the interference dip due to an increase in the refractive index of the coated cladding of the fiber. A quite higher response/recovery time of 89 s/97 s was noted by this sensor. The ZnO embedded in the PMMA matrix was reported for H_2S sensing (Kitture et al. 2017). The sensing approach was based on Fabry-Perot interferometric cavity. Three types of probes were prepared and employed for H_2S sensing in 1 ppm–5 ppm. The sensor exhibited a responsivity of 1950 pm/ppm and a decay time of 27 s was noted. However, due to PMMA material, the sensor response may be altered at a high working temperature.

5.4.2 H_2 Sensor

H_2 is tremendously flammable and explosive. It has a low explosion limit (~4%) and a wide explosion range (4–75%) in the air. Therefore, the detection system must be immune to charge and heat and thus, fiber-optic sensor (FOS) is highly suitable for the detection of H_2 gas. Lots of H_2 sensors have been fabricated based on FOS approach.

Mostly, for H_2 gas, extensive use of noble metals such as Pt, Pd, Ag, and Au has been observed. The fact is that the metals act as catalysts and own high hydrogen chemisorption properties. The typical reaction kinetics involved in the detection of H_2 detection is given below.

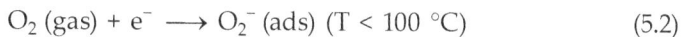

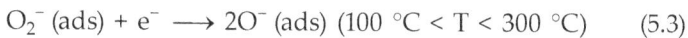

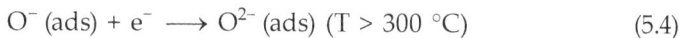

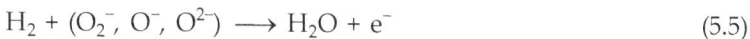

$$O_2 \text{ (gas)} \longrightarrow O_2 \text{ (ads)} \tag{5.1}$$

$$O_2 \text{ (gas)} + e^- \longrightarrow O_2^- \text{ (ads) (T < 100 °C)} \tag{5.2}$$

$$O_2^- \text{ (ads)} + e^- \longrightarrow 2O^- \text{ (ads) (100 °C < T < 300 °C)} \tag{5.3}$$

$$O^- \text{ (ads)} + e^- \longrightarrow O^{2-} \text{ (ads) (T > 300 °C)} \tag{5.4}$$

$$H_2 + (O_2^-, O^-, O^{2-}) \longrightarrow H_2O + e^- \tag{5.5}$$

The metallic Pd is used for H_2 sensing as it is capable to absorb H_2 up to 900 times its volume and can dissociate into two hydrogen atoms which then form a palladium hydride. Thus, this changes the optical, and electrical properties of the materials and hence, is measured accordingly. The Pd-WO_3 combination has also been widely used for H_2 sensing (Zhang et al. 2016). As the WO_3 changed its color in presence of H_2 gas it easily modulated the refractive index of the material. A modal interferometer utilized with Pd-WO_3 to 0–10 000 µl/l concentration range of H_2 gas was fabricated. The sensor showed a shift of 0.109 pm/(µl/l) with a detection limit of 440.367 µl/l. But it exhibited a longer rising time of around <33 min which could be a demerit of this system. The tapered silica MMF was also utilized for H_2 in 100–2000 ppm (Sun et al. 2017). The low detection of 100 ppm with longer response/recovery time was detected. The response of the device was also measured by mixing different concentrations of H_2, CH_4, and CO in a balanced N_2 environment.

5.4.3 CH_4 Sensor

Methane is colorless, flammable and explosive in nature, and therefore, regular monitoring via a safe and reliable sensory system is very important. The exposure limit of CH_4 gas is 4.3%–16.2%. Some CH_4 sensor development is provided below.

It is observed that Cryptophane A is widely chosen for CH_4 sensing because Cryptophane A has a large absorption ability and is mostly inert to other gases. Polycarbonate/ Cryptophane A coated on LPG was designed for CH_4 detection (Yang et al. 2015b). The overlay coating thickness of 530 was deposited by using a dip coating technique. The sensor showed a significant wavelength shift of 2.5 nm/% and a low detection limit of 0.2% (v/v) over CH_4 concentration of 0–3.5% and RI sensitivity could reach up to 3.56×10^3 nm/RIU. The incorporated SnO_2 into graphene nanocomposite was explored for CH_4 sensing (Zhang et al. 2017b). The intensity-based technique is quite simple but exhibited a lower sensitivity. The methane concentration varied from 0% to 55% and the corresponding light intensity was measured. A linear relation was observed between the light intensity and CH_4 concentration. However, the cross sensitivity to other gases is not studied.

5.4.4 NH_3 Sensor

Ammonia is a colorless gas at room temperature and irritating to the eyes, nose, throat, and respiratory tract and can even lead to death at high exposure levels. The exposure limit of 50 pm is set by the Occupational

Safety and Health Administration (OSHA), USA. Due to potential health hazards to the human being, it is required to continuously monitor the NH_3 in the atmosphere. Some NH_3 sensors developed are discussed below.

The ZnO material is layered via a dipping method on a clad removed part of the multimode fiber (Devendiran and Sastikumar 2017). The film thickness observed was around 250 μm and it was exposed to 0–500 ppm NH_3 and the corresponding change in intensity was recorded with a vapor pressure of the gas. This absorption-based method displayed a longer response/recovery time of around 11 min/8 min. However, the sensor was also tested for different volatile compounds as well. Another NH_3 sensor based on polymer and metal coating was fabricated by using Ag/PVP/PVA material on the unclad part of the POF (Rithesh Raj et al. 2015). This sensor was also based on the intensity variation of the spectra with the NH_3 concentration. The concentration of Ag nanoparticles in the PVP/PVA hybrid was changed from 1.6, 3.3, and 6.6% and the corresponding response measured to 0–500 ppm NH_3 further achieved a responsivity around 0.88 counts/ppm. However, the temporal characteristics of the sensor were very weak which exhibited around 90 min each.

5.4.5 CO_2 and NO_2 Sensor

Fiber grating was also employed for the detection of CO_2 sensing (Wu et al. 2016b). The intensity-based LPG sensor was explored in the 3–15% CO_2 concentration and showed a response of 0.0513 dB/%. However, the response and recovery time was a little higher ~6 min/~20 min, respectively. NO_2 sensor based on FBG was also tested and observed good sensing results. The rGO acted as a sensing material and was coated onto the etched part of the single mode photosensitive optical fiber (Sridevi et al. 2016). The rGO was coated by using a dip coating method. The modified part was exposed to NO_2 concentration of 0.5–3 ppm and exhibited a good response of 34.3 pm at 3 ppm NO_2 gas with LOD ~0.5 ppm. But the rising and decay time observed was longer at around 12.3 min/3.8 min, respectively, and also the sensor humidity and interfering effect from other gases were not examined.

5.4.6   VOCs Sensor

In the recent past, numerous fiber optics vapor sensors have been demonstrated. Various materials including metal oxides, graphene, and polymeric materials have been employed. The simple technique like dip coating, and drop casting are utilized in the fabrication of cheap and flexible FOS sensors.

A polarization maintaining microfiber was explored for the measurement of ethanol as well. The tapered section of the microfiber was coated with GO and exposed to ethanol (Zhang et al. 2017a). A highly sensitive system was developed and obtained a sensitivity of 0.138 nm/ppm to 0–80 ppm ethanol. The system detected a minimum ethyl alcohol of around 1.6 ppm, but showed a little longer response time of 6.1 min. Nanopatterned ethylene sensor was fabricated by using TiO_2 and GO nanomaterials (Tabassum et al. 2017). The device was sensitive and showed good a sensitivity of 0.92 pm/ppm to 0–909 ppm ethylene concentration. The periodic nanostructures formed at the fiber tip were providing light trapping at the nanoscale and thus, interacted effectively. This type of sensor has great potential in the near future due to its simplicity and cost-effectiveness.

5.5 QUARTZ CRYSTAL MICROBALANCE-BASED GAS SENSORS

QCM can detect a minute change in the mass of the gas or vapor molecules. By using this principle many gas sensors have been fabricated. Additionally, this device offers stability and room temperature operation. The QCM based gas sensors are shown in Figure 5.5. Table 5.13 depicts the summary of some QCM-based gas sensors.

The MOX nanostructures coupled with QCM were also investigated for various gases and vapor sensing. In short, the V_2O_5 nanostructures were studied toward NH_3 sensing (Berouaken et al. 2020). The rise in Δf stimulated by the increase in mass on the QCM with respect to the adsorption of NH_3 molecules was calculated by using the Sauerbrey equation and it was observed that the device showed a sensitivity of 2.05 Hz/ppm with quick rise and decay time and excellent stability, reproducibility with linearity. The 2D materials were also utilized for monitoring toxic gas and vapors. In this, a WS_2 thin film was coated by using the magnetron sputtering technique onto an AT-cut QCM device at room temperature. The frequency shift with time response was measured with all gases like acetone, toluene, xylene, isopropyl-alcohol, HCN, H_2S, NH_3, DMMP, and 2-Chloroethyl ethyl sulfide (Alev et al. 2022). By using these device, nine different gasses were tested. The interference magnitude of the WS_2/QCM sensor was around 2 Hz and interestingly, the without coated sensor did not display any response. Among WS_2 sensors, the highest response was observed with DMMP followed by H_2S, NH_3, acetone, CEES, HCN, xylene, IPA, and toluene. The sensor's long-term stability of over 2 months was tested. The highest sensing response to DMMP was attributed due to the double bond P=O and –OCH groups

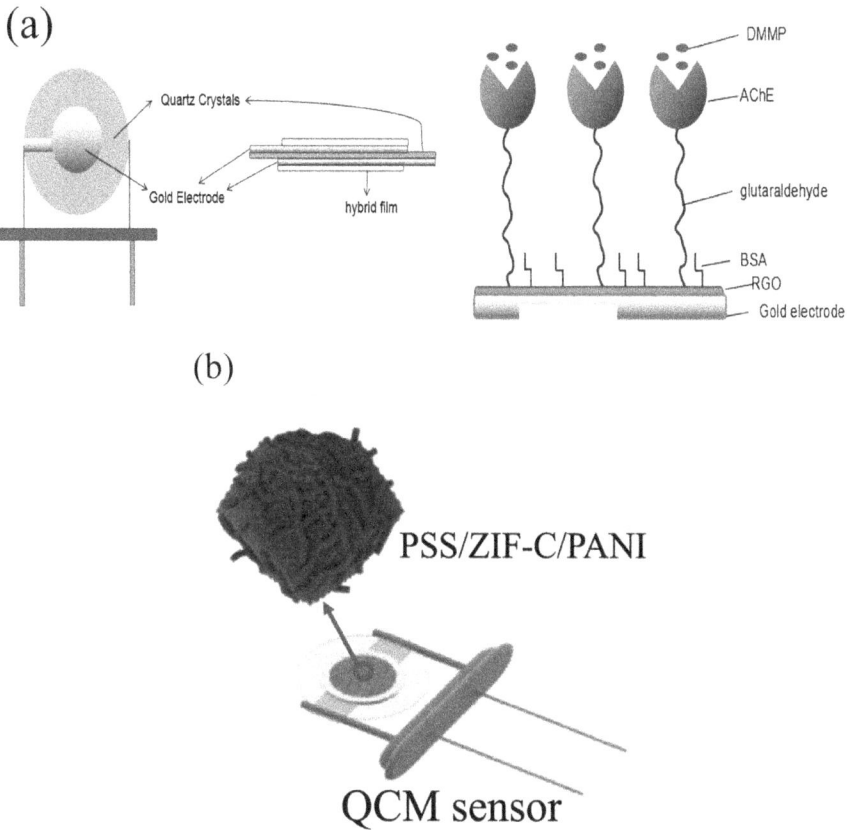

Figure 5.5 The quartz crystal microbalance-based gas sensors. (a) Acetylcholinesterase-reduced graphene oxide hybrid films hybrid film for DMMP sensing, [Reproduced with permission (Tang et al. 2016). Copyright 2016, Elsevier]. (b) PSS doped ZIF-8-derived porous carbon/polyaniline hybrid film for NH_3 sensing. [Reproduced with permission (Zhang et al. 2022c). Copyright 2022, Elsevier].

of DMMP, which strongly binds with the active sites of the WS_2 material. The LOD of the device was measured up to 5 ppb. The heterojunction of MOX was also employed for the QCM-based gas sensing applications. For instance, the GO/Cu_2O was coated onto the QCM chip by a layer-by-layer assembly method (Chen et al. 2018). The sensor showed better selectivity, stability, and selectivity to TMA gas with a sensitivity of 8.91 Hz/ppm and low LOD of 0.23 ppm. However, the sensor was influenced by humidity. The functionalization technique was also very effective for the gas sensing applications. Along with these

Table 5.13 Summary of a few quartz crystal microbalance-based gas sensors based on nanomaterials

Materials used	Gas analyte	QCM fundamental frequency (MHz)	Minimum Concentration detected (ppm)	Sensor sensitivity (Hz/ppm)	Response/ Recovery time (s)	References
rGO/AChE	Dimethyl methylphosphonate	8	10 mg/m^3	ca. 10 Hz (10 mg/m^3)	–	(Tang et al. 2016)
PSS/ZIF-C/PANI	NH_3	8	5	5.67	28/17 @10 ppm	(Zhang et al. 2022c)
V_2O_5	NH_3	–	9	2.05	–	(Berouaken et al. 2020)
CN-COOH	NH_3	10	50	–	24/119 @50 ppm	(Wang and Song 2020)
rGO/SnO$_2$/AgNPs	NO_2	10	2.05 mg/m^3	491 Hz (2.05 mg/m^3)	–	(Qi et al. 2016)
WS$_2$	Acetone	10	0.95	2.61	–	(Alev et al. 2022)
	H_2S	10	0.875	7.25	–	
	HCN	10	1.06	1.91	–	
	NH_3	10	1.45	3.57	–	
	DMMP	10	0.175	37.38	–	
	Toluene	10	240	0.06	–	
	Xylene	10	80	0.35	–	
V$_2$O$_5$/ZnO	Chlorobenzene	10	100	~0.13	–	(Kösemen et al. 2017)
	n-Heptane	10	390	~0.03	–	
	Tetrachloroethylene	10	150	~0.08	–	
	O-Xylene	10	40	~0.31	–	
	Toluene	10	240	0.06	–	
	Ethylbenzene	10	70	~0.18	–	
	Acetonitrile	10	950	~0.01	–	
	Methanol	10	1000	~0.01	–	
	Isopropanol	10	130	~0.08	–	
	Ethyl acetate	10	900	~0.025	–	
	Triethylamine	10	720	~0.03	–	

(Contd.)

Materials used	Gas analyte	QCM fundamental frequency (MHz)	Minimum Concentration detected (ppm)	Sensor sensitivity (Hz/ppm)	Response/ Recovery time (s)	References
	Ethanol	10	12.2	1.811	–	(Öztürk et al. 2016)
	Methanol	10	–	0.435	–	
	Isopropanol	10	–	1.22	–	
1.5% Pd doped ZnO	Xylene	10	–	3.296	–	
	Toluene	10	11.6	0.808	–	
	Acetone	10	76	0.159	–	
	Chloroform	10	76	0.337	–	
Silver(I)/ polymer composite	Ethylene	–	1	50.748	–	(Tolentino et al. 2018)
GO/Cu$_2$O	Trimethylamine (TMA)	–	0.6	8.91	20	(Chen et al. 2018)
Amino-functionalized-GO	Aldehyde	9	45	8.06	14/150	(Chen et al. 2020)
ZIF-8	CO$_2$	9	–	2.18 Hz/ vol%	10	(Devkota et al. 2018)
MIL-101(Cr)	Pyridine	10	5	2.793	7.67/61	(Haghighi and Zeinali 2019)
	Methanol	10	25	0.306	7.67/20.33	
	Ethanol	10	10	0.429	10/38.33	
	2-Propanol	10	5	0.648	7/29.67	
	Acetone	10	25	0.114	6.33/25.33	
	Chloroform	10	5	0.524	3.67/42.33	
	Dichloromethane	10	25	0.155	8/30.33	
	n-Hexane	10 MHz	25	0.164	11/24.67	
	THF	10 MHz	25	0.205	5.67/21	

$\Delta F = -\dfrac{2F^2}{\sqrt{r_q m_q}}\dfrac{\Delta M}{A}$, where m_q and r_q are the shear modulus (g cm^{-1} s^{-2}) and density of quartz crystal (g cm^{-3}).

the hydrophobic nature could add the water-resistant property, and therefore, will increase the durability of the sensor. By considering this, the amino-functionalized graphene oxide nanocomposite was coated onto the QCM chip and investigated to detect the aldehydes in the fish fillets (Chen et al. 2020). The sensor showed a good response to hexanal, octanal, and nonanal gases. However, the frequency shift was small, and therefore, needs to differentiate further. At present, MOFs are of immense interest in gas detection because of their miscellaneous structures with even pores, huge surface area, high gas adsorption, and good thermal stability. The QCM-based resonator was coated with a ZIF-8 on a QCM chip by a solution-based dip coating method (Haghighi and Zeinali 2019). The decrease in resonant frequency @4250 Hz, which corresponded to the mass density was observed. When the sensor was exposed to CO_2 gas, a decrease in resonant frequency was experienced, and therefore, the sensitivity of the sensor was calculated at around 2.18 Hz/vol%. Despite, good sensor stability, the humidity and cross-selectivity to other gases were not analyzed.

5.6   CONCLUSION AND OUTLOOK

In summary, the recent progress in the development of gas sensing based on numerous advanced and reliable methods is presented. At present, these sensing methods offer high sensitivity, selectivity, material adaptability, flexibility, and reliability. The integration of various materials such as metal oxides, 2D material, polymers, and their composites with these platforms for different gas sensing applications is reviewed. Various gases such as H_2S, H_2, CO, NO_2, CO_2, NH_3, CH_4, SO_2, and a few VOCs such as ethanol, methanol, acetone, chlorobenzene, toluene, etc. are detected. These methods are very effective and offer great gas sensing performance and applicability towards environmental and industrial applications. The key points including the role of each sensing method and the effect of material property in gas sensing performance are addressed. The limitations of each method along with the possible scope for improvement are also presented.

The key parameters in gas sensing are analyzed as given below:

1. Generally, the gas sensing characteristics are mainly dependent upon the material's properties. It was observed that the size, shape, morphology, porosity, and surface area play a crucial role in deciding the gas sensing performance.

2. The highly porous materials exhibit a high sensing response and fast temporal response characteristics. As the gas can easily interact with the inner walls of the materials and it offers a high selectivity.

3. It was observed that the nanocomposite materials are mostly used for gas sensing applications. It not only increases the response and selectivity but also decreases the working temperature of the device. The nanocomposites possess heterojunction and thus,

many active sites at the depletion layer are available for the gas interaction and thus, enhance the sensor response along with the selectivity. Sometimes, the BET specific surface area for nanocomposite may be lower due to the blocking of pores (it might occur with metals or other small nanoparticles coated materials).

4. Numerous room temperature gas sensors are reported and exhibited a good sensing response. However, the response is not up to the mark and can improve further by adopting strategies like noble metal modified materials; metal ion doped, composites with carbon nanomaterials, and 2D materials.

5. Another strategy of metal coating was effectively explored for gas detection. The general advantages of metal coating are the catalytic effect, lowering the working temperature, and improvement in rise and decay time along with selectivity.

6. The metal ion doping in materials can effectively enhance the sensor response as it creates lot many active sites and thus, more formation of oxygen species for interaction.

7. The carbon-based materials are widely investigated in gas sensing because of their large conductivity and large adsorption capability and thus, can be used for fast response.

8. Light-based techniques are used for gas sensing. It improved the response, time characteristics, and working temperature. However, the adaptability, power consumption, and portability of the sensor also need to be considered for on-site monitoring applications.

9. The polymer-based gas sensors are attracting great attention due to their flexibility and room temperature ability.

10. Chemiresistive-based gas sensors are very efficient due to their simple and low-cost fabrication. Several room temperature gas sensors have been developed but most of them offer a low sensitivity and selectivity. Generally, the metal oxides are sensitive at higher temperatures and also the humidity impact at low temperature could be the reason for the degradation of the sensor response.

11. The fiber optics sensors are extensively used in monitoring gases and vapors because of their simple, fast response, ability to integrate with any material, and on-line and multiplexing capability. However, these sensors exhibited certain limitations such as the response is dependent upon the coating thickness. The smaller thickness exhibits a larger response and fast response and recovery time therefore, controlling the thickness is very

important. Very few sensors are reported for the detection of the mixture, which is important for practical applications.

12. The electrochemical sensors have been widely adopted for gas sensing and have shown good gas sensing results. However, the appropriate materials are not accessible for high temperature operation. Also, the reaction kinetics is dependent upon many factors such as, processing conditions and microstructural changes during operation, which might change sensor response over time. The cross-selectivity is also an issue. The amperometry-based gas sensor is widely reported. However, the generated current is small and there is a need to avoid the interfering current in the output.

13. In recent years, an inter-digital capacitive-based approach has considerably increased for gas sensing applications. This structure is highly flexible, low-cost, and easy to prepare. Generally, the change in capacitance is based on alteration in the dielectric constant of the sensing material in presence of gas molecules. However, the change in dielectric constant for most of the gas molecules is nearly the same and therefore, this approach is not much sensitive for gas sensing applications. So, there is a scope to develop a capacitive sensor by using polymeric-based nanocomposites materials which changes thickness due to swelling.

14. Field effect transistor is the also mostly used approach for gas sensing due to their flexibility, small size, high response, and speedy nature. There is a scope to improve the gas sensing performance further. The proper functionalization and the right selection of gate material is necessary to enhance the selectivity. MOXs operated FET devices worked well at higher temperature and thus, need to focus on strategies like doping of metal ions, preparing carbon and polymeric-based nanocomposites. Black phosperous showed good sensitivity but is prone to oxidation and moisture. Most of the FETs show a longer response and recovery time and so, other strategies like the use of a heating source can be implemented. The use of polymeric-based substrate is useful for the development of flexible and stretchable devices.

15. Quartz crystal microbalance is a very attractive method for gas detection owing to its high sensitivity, easy preparation, and low-cost setup. A lot of graphene and its derivatives-based gas sensors have been explored and offered a high sensing performance. However, the QCM is based on only graphene and show less sensitivity and slow recovery, and therefore, there is a need to apply functionalization techniques to improve the sensing performance. Numerous materials like metal oxides,

2D materials and polymeric materials can be combined and explored for gas sensing. In the case of QCM, reproducibility is an issue, and therefore, needs to be handled properly. As QCM is a highly sensitive method and so, the interfering effect of humidity on gas sensing material also needs to be studied.

There is always a scope to improve the gas sensing performance by adopting the following possible ways.

1. The sensor selectivity can be improved by incorporating metal and designing nanocomposites with a suitable material.

2. Most of the sensors show poor long-term stability. There are certain influencing factors like a humid environment and temperature which can affect device durability. The controlled material morphology of the materials, use of an external heater or light illumination, and modeling can be used to overcome this issue.

3. Selectivity is a crucial factor from a gas sensing point of view. It can be improved by modulating the sensing materials, specific gas sensing analysis based on chemical adsorption or physical adsorption for sensing materials, usage of filters (sorption, size-selective and catalytic), controlling temperature and humidity, utilization of a gas sensor in a harsh practical environment, and implementation of pattern recognition techniques, etc.

4. Numerous materials and composites with different sizes, shapes and morphology are used for gas sensing. However, it is not addressed that the same material with alike sizes and morphologies shows dissimilar sensing properties.

5. The relationship between the enhanced performance and modified properties should be more precise for a better understanding of the electronic or chemical effects.

6. A lot of studies has been reported on 2D/MOX composites. However, the proper sensing mechanism of gas interaction is still underway.

7. In the case of the heterostructure, the detection of gases depends upon the fermi energy difference and altering the depletion width. However, along with these, an optimized composition ratio of each material should also be considered.

8. With advanced nanotechnology, various deposition techniques, wide synthesis methods, and different morphology of nanomaterials, a gas sensor based on the multivariable approach can be fabricated. In this approach, a material having different mechanisms and many transducers (electrical, optical, electrochemical, capacitive) could be produced.

REFERENCES

Abokifa, Ahmed A., Kelsey Haddad, Baranidharan Raman, John Fortner and Pratim Biswas. 2021. Room temperature gas sensing mechanism of SnO_2 towards chloroform: comparing first principles calculations with sensing experiments. Applied Surface Science 554: 149603. doi:https://doi.org/10.1016/j.apsusc.2021.149603.

Ahmad, Khursheed, Akbar Mohammad and Shaikh M. Mobin. 2017. Hydrothermally grown α-MnO_2 nanorods as highly efficient low cost counter-electrode material for dye-sensitized solar cells and electrochemical sensing applications. Electrochimica Acta 252: 549–557. doi:https://doi.org/10.1016/j.electacta.2017.09.010.

Alev, Onur, Okan Özdemir, Eda Goldenberg, Leyla Çolakerol Arslan, Serkan Büyükköse and Zafer Ziya Öztürk. 2022. WS_2 thin film based quartz crystal microbalance gas sensor for dimethyl methylphosphonate detection at room temperature. Thin Solid Films 745: 139097. doi:https://doi.org/10.1016/j.tsf.2022.139097.

Amruth, K., K.M. Abhirami, S. Sankar and M.T. Ramesan. 2022. Synthesis, characterization, dielectric properties and gas sensing application of polythiophene/chitosan nanocomposites. Inorganic Chemistry Communications 136: 109184. doi:https://doi.org/10.1016/j.inoche.2021.109184.

Andrés, Miguel A., Mani Teja Vijjapu, Sandeep G. Surya, Osama Shekhah, Khaled Nabil Salama, Christian Serre, Mohamed Eddaoudi, et al. 2020. Methanol and humidity capacitive sensors based on thin films of mof nanoparticles. ACS Applied Materials & Interfaces 12(3): 4155–4162. American Chemical Society. doi:10.1021/acsami.9b20763.

Bai, Xue, Zhuo Liu, He Lv, Junkun Chen, Mawaz Khan, Jue Wang, Baihe Sun, et al. 2022. N-doped three-dimensional needle-like CoS_2 bridge connection Co_3O_4 core–shell structure as high-efficiency room temperature NO_2 gas sensor. Journal of Hazardous Materials 423: 127120. doi:https://doi.org/10.1016/j.jhazmat.2021.127120.

Berouaken, Malika, Lamia Talbi, Chafiaa Yaddadene, Mohamed Maoudj, Hamid Menari, Rezak Alkama and Noureddine Gabouze. 2020. Room temperature ammonia gas sensor based on V_2O_5 nanoplatelets/quartz crystal microbalance. Applied Physics A 126(12): 949. doi:10.1007/s00339-020-04129-6.

Bhardwaj, Radha and Arnab Hazra. 2022. Realization of ppb-level acetone detection using noble metals (Au, Pd, Pt) nanoparticles loaded GO FET sensors with simultaneous back-gate effect. Microelectronic Engineering 256: 111719. doi:https://doi.org/10.1016/j.mee.2022.111719.

Bo, Zheng, Mu Yuan, Shun Mao, Xia Chen, Jianhua Yan and Kefa Cen. 2018. Decoration of vertical graphene with tin dioxide nanoparticles for highly sensitive room temperature formaldehyde sensing. Sensors and Actuators B: Chemical 256: 1011–1020. doi:https://doi.org/10.1016/j.snb.2017.10.043.

Cao, Anping, Meixia Shan, Laura Paltrinieri, Wiel H. Evers, Liangyong Chu, Lukasz Poltorak, Johan H Klootwijk, et al. 2018. Enhanced vapour sensing using silicon nanowire devices coated with Pt nanoparticle functionalized porous organic frameworks. Nanoscale 10(15): 6884–6891. The Royal Society of Chemistry. doi:10.1039/C7NR07745A.

Cao, PeiJiang, YongZhi Cai, Dnyandeo Pawar, S.T. Navale, Ch.N. Rao, Shun Han, WangYin Xu, et al. 2020. Down to ppb level NO₂ detection by ZnO/rGO heterojunction based chemiresistive sensors. Chemical Engineering Journal 401: 125491. doi:https://doi.org/10.1016/j.cej.2020.125491.

Cao, PeiJiang, YongZhi Cai, Dnyandeo Pawar, Shun Han, WangYing Xu, Ming Fang, XinKe Liu, et al. 2022. Au@ZnO/rGO nanocomposite-based ultralow detection limit highly sensitive and selective NO₂ gas sensor. Journal of Materials Chemistry C 10(11): 4295–4305. The Royal Society of Chemistry. doi:10.1039/D1TC05835E.

Chakraborty, Urmila, Preeti Garg, Gaurav Bhanjana, Gurpreet Kaur, Ajeet Kaushik and Ganga Ram Chaudhary. 2022. Spherical silver oxide nanoparticles for fabrication of electrochemical sensor for efficient 4-nitrotoluene detection and assessment of their antimicrobial activity. Science of the Total Environment 808: 152179. doi:https://doi.org/10.1016/j.scitotenv.2021.152179.

Chen, Wei, Fanfei Deng, Min Xu, Jun Wang, Zhenbo Wei and Yongwei Wang. 2018. GO/Cu₂O nanocomposite based QCM gas sensor for trimethylamine detection under low concentrations. Sensors and Actuators B: Chemical 273: 498–504. doi:https://doi.org/10.1016/j.snb.2018.06.062.

Chen, Wei, Zhenhe Wang, Shuang Gu, Jun Wang, Yongwei Wang and Zhenbo Wei. 2020. Hydrophobic amino-functionalized graphene oxide nanocomposite for aldehydes detection in fish fillets. Sensors and Actuators B: Chemical 306: 127579. doi:https://doi.org/10.1016/j.snb.2019.127579.

Cheng, Chu, Jie Zou, Yucun Zhou, Zi Wang, Han Jin, Guangzhong Xie and Jiawen Jian. 2021a. Fabrication and electrochemical property of La₀.₈Sr₀.₂MnO₃ and (ZrO₂)0.92(Y₂O₃)0.08 interface for trace alcohols sensor. Sensors and Actuators B: Chemical 331: 129421. doi:https://doi.org/10.1016/j.snb.2020.129421.

Cheng, Lingli, Yongchao He, Maozhi Gong, Xinhua He, Zhukai Ning, Hongchuan Yu and Zheng Jiao. 2021b. MOF-derived synthesis of Co₃O₄ nanospheres with rich oxygen vacancies for long-term stable and highly selective n-butanol sensing performance. Journal of Alloys and Compounds 857: 158205. doi:https://doi.org/10.1016/j.jallcom.2020.158205.

Chernikova, Valeriya, Omar Yassine, Osama Shekhah, Mohamed Eddaoudi and Khaled N. Salama. 2018. Highly sensitive and selective SO₂ MOF sensor: the integration of MFM-300 MOF as a sensitive layer on a capacitive interdigitated electrode. Journal of Materials Chemistry A 6(14): 5550–5554. The Royal Society of Chemistry. doi:10.1039/C7TA10538J.

Cho, Soo-Yeon, Kyeong Min Cho, Sanggyu Chong, Kangho Park, Sungtak Kim, Hohyung Kang, Seon Joon Kim, et al. 2018. Rational design of aminopolymer for selective discrimination of acidic air pollutants. ACS Sensors 3(7): 1329–1337. American Chemical Society. doi:10.1021/acssensors.8b00247.

Chua, Wen Hong, Mohd Hanif Yaacob, Chou Yong Tan and Boon Hoong Ong. 2021. Chemical bath deposition of H-MoO₃ on optical fibre as room-temperature ammonia gas sensor. Ceramics International 47(23): 32828–32836. doi:https://doi.org/10.1016/j.ceramint.2021.08.179.

Cui, Guangliang, Pinhua Zhang, Li Chen, Xiaoli Wang, Jianfu Li, Changmin shi and Dongchao Wang. 2017. Highly sensitive H_2S sensors based on Cu_2O/Co_3O_4 nano/microstructure heteroarrays at and below room temperature. Scientific Reports 7(1): 43887. doi:10.1038/srep43887.

Das, Arindam and Dipankar Panda. 2019. SnO_2 tailored by CuO for improved CH_4 sensing at low temperature. Physica Status Solidi (B) 256(5): 1800296. John Wiley & Sons, Ltd. doi:https://doi.org/10.1002/pssb.201800296.

Devendiran, S. and D. Sastikumar. 2017. Gas sensing based on detection of light radiation from a region of modified cladding (Nanocrystalline ZnO) of an optical fiber. Optics & Laser Technology 89: 186–191. doi:https://doi.org/10.1016/j.optlastec.2016.10.013.

Devkota, Jagannath, Ki-Joong Kim, Paul R. Ohodnicki, Jeffrey T. Culp, David W. Greve and Jonathan W. Lekse. 2018. Zeolitic imidazolate framework-coated acoustic sensors for room temperature detection of carbon dioxide and methane. Nanoscale 10(17): 8075–8087. The Royal Society of Chemistry. doi:10.1039/C7NR09536H.

Doblinger, Simon, Catherine E. Hay, Liliana C. Tomé, David Mecerreyes and Debbie S. Silvester. 2022. Ionic liquid/poly(ionic liquid) membranes as non-flowing, conductive materials for electrochemical gas sensing. Analytica Chimica Acta 1195: 339414. doi:https://doi.org/10.1016/j.aca.2021.339414.

Drmosh, Q.A., Z.H. Yamani, A.K. Mohamedkhair, A.H.Y. Hendi, M.K. Hossain and Ahmed Ibrahim. 2018. Gold nanoparticles incorporated SnO_2 thin film: highly responsive and selective detection of NO_2 at room temperature. Materials Letters 214: 283–286. doi:https://doi.org/10.1016/j.matlet.2017.12.013.

Drmosh, Q.A., A.H. Hendi, M.K. Hossain, Z.H. Yamani, R.A. Moqbel, Abdo Hezam and M.A. Gondal. 2019. UV-activated gold decorated rGO/ZnO heterostructured nanocomposite sensor for efficient room temperature H_2 detection. Sensors and Actuators B: Chemical 290: 666–675. doi:https://doi.org/10.1016/j.snb.2019.03.077.

Dunst, Katarzyna, Jakub Karczewski and Piotr Jasiński. 2017. Nitrogen dioxide sensing properties of PEDOT polymer films. Sensors and Actuators B: Chemical 247: 108–113. doi:https://doi.org/10.1016/j.snb.2017.03.003.

Evans, Gwyn P., Michael J. Powell, Ian D. Johnson, Dougal P. Howard, Dustin Bauer, Jawwad A. Darr and Ivan P. Parkin. 2018. Room temperature vanadium dioxide–carbon nanotube gas sensors made via continuous hydrothermal flow synthesis. Sensors and Actuators B: Chemical 255: 1119–1129. doi:https://doi.org/10.1016/j.snb.2017.07.152.

Faruk Hossain, Md, Stephanie McCracken and Gymama Slaughter. 2021. Electrochemical laser induced graphene-based oxygen sensor. Journal of Electroanalytical Chemistry 899: 115690. doi:https://doi.org/10.1016/j.jelechem.2021.115690.

Fu, Huifen, Changliang Hou, Fubo Gu, Dongmei Han and Zhihua Wang. 2017. Facile preparation of rod-like Au/In_2O_3 nanocomposites exhibiting high response to CO at room temperature. Sensors and Actuators B: Chemical 243: 516–524. doi:https://doi.org/10.1016/j.snb.2016.11.162.

Fu, Haiwei, Qiqi Wang, Jijun Ding, Yi Zhu, Min Zhang, Chong Yang and Shuai Wang. 2020. Fe_2O_3 nanotube coating micro-fiber interferometer for ammonia detection. Sensors and Actuators B: Chemical 303: 127186. doi:https://doi.org/10.1016/j.snb.2019.127186.

Fu, Haiwei, Yongtao You, Shuai Wang and Huimin Chang. 2022. SnO_2 nanomaterial coating micro-fiber interferometer for ammonia concentration measurement. Optical Fiber Technology 68: 102819. doi:https://doi.org/10.1016/j.yofte.2022.102819.

Fuśnik, Łukasz, Bartłomiej Szafraniak, Anna Paleczek, Dominik Grochala and Artur Rydosz. 2022. A review of gas measurement set-ups. Sensors. doi:10.3390/s22072557.

Gao, Hongyu, Jie Guo, Yiwen Li, Changlin Xie, Xiao Li, Long Liu, Yi Chen, et al. 2019. Highly selective and sensitive xylene gas sensor fabricated from NiO/$NiCr_2O_4$ p-p Nanoparticles. Sensors and Actuators B: Chemical 284: 305–315. doi:https://doi.org/10.1016/j.snb.2018.12.152.

Gao, Huijun, Yuzhen Ma, Peng Song, Zhongxi Yang and Qi Wang. 2021. Cu-doped Fe_2O_3 porous spindles derived from metal-organic frameworks with enhanced sensitivity to triethylamine. Materials Science in Semiconductor Processing 123: 105510. doi:https://doi.org/10.1016/j.mssp.2020.105510.

Gu, Wenxiu, Wangwang Zheng, Han Liu and Yuan Zhao. 2021. Electroactive Cu_2O nanocubes engineered electrochemical sensor for H_2S detection. Analytica Chimica Acta 1150: 338216. doi:https://doi.org/10.1016/j.aca.2021.338216.

Guo, Yicheng, Xianqing Tian, Xinfeng Wang and Jie Sun. 2019. Fe_2O_3 nanomaterials derived from prussian blue with excellent H_2S sensing properties. Sensors and Actuators B: Chemical 293: 136–143. doi:https://doi.org/10.1016/j.snb.2019.04.027.

Ha, Nguyen Hai, Dao Duc Thinh, Nguyen Thanh Huong, Nguyen Huy Phuong, Phan Duy Thach and Hoang Si Hong. 2018. Fast response of carbon monoxide gas sensors using a highly porous network of ZnO nanoparticles decorated on 3D reduced graphene oxide. Applied Surface Science 434: 1048–1054. doi:https://doi.org/10.1016/j.apsusc.2017.11.047.

Haghighi, Elahe and Sedigheh Zeinali. 2019. Nanoporous MIL-101(Cr) as a sensing layer coated on a quartz crystal microbalance (QCM) nanosensor to detect volatile organic compounds (VOCs). RSC Advances 9(42): 24460–24470. The Royal Society of Chemistry. doi:10.1039/C9RA04152D.

Haridas, Vijayasree, A. Sukhananazerin, J. Mary Sneha, Biji Pullithadathil and Binitha Narayanan. 2020. α-Fe_2O_3 loaded less-defective graphene sheets as chemiresistive gas sensor for selective sensing of NH_3. Applied Surface Science 517: 146158. doi:https://doi.org/10.1016/j.apsusc.2020.146158.

Hong, Yoonki, Chang-Hee Kim, Jongmin Shin, Kyoung Yeon Kim, Jun Shik Kim, Cheol Seong Hwang and Jong-Ho Lee. 2016. Highly selective ZnO gas sensor based on MOSFET having a horizontal floating-gate. Sensors and Actuators B: Chemical 232: 653–659. doi:https://doi.org/10.1016/j.snb.2016.04.010.

Hosseini, Z.S., A. Iraji zad and A. Mortezaali. 2015. Room temperature H_2S gas sensor based on rather aligned ZnO nanorods with flower-like structures. Sensors and Actuators B: Chemical 207: 865–871. doi:https://doi.org/10.1016/j.snb.2014.10.085.

Hsu, Cheng-Liang, Bo-Yu Jhang, Cheng Kao and Ting-Jen Hsueh. 2018. UV-illumination and au-nanoparticles enhanced gas sensing of p-type Na-doped ZnO nanowires operating at room temperature. Sensors and Actuators B: Chemical 274: 565–574. doi:https://doi.org/10.1016/j.snb.2018.08.016.

Huang, Jinyu, Dongting Jiang, Jiaxi Zhou, Jiexiong Ye, Yiling Sun, Xuejin Li, Youfu Geng, et al. 2021. Visible light-activated room temperature NH_3 sensor base on CuPc-loaded ZnO nanorods. Sensors and Actuators B: Chemical 327: 128911. doi:https://doi.org/10.1016/j.snb.2020.128911.

Ibrahim, Siti Azlida, Norizah Abdul Rahman, Mohd Hanif Yaacob, Muhammad Hafiz Abu Bakar, Fatimah Syahidah Mohamad and Nor Akmar Mohd Yahya. 2022. Self-assembled polyaniline nanostructures in situ deposited on silica optical fibers for ammonia gas sensing. Synthetic Metals 283: 116962. doi:https://doi.org/10.1016/j.synthmet.2021.116962.

Ingle, Nikesh, Savita Mane, Pasha Sayyad, Gajanan Bodkhe, Theeazen AL-Gahouari, Manasi Mahadik, Sumedh Shirsat, et al. 2020. Sulfur dioxide (SO_2) detection using composite of nickel benzene carboxylic (Ni_3BTC_2) and OH-functionalized single walled carbon nanotubes (OH-SWNTs). Frontiers in Materials. https://www.frontiersin.org/article/10.3389/fmats.2020.00093.

Jang, Ji-Soo, Hayoung Yu, Seon-Jin Choi, Won-Tae Koo, Jiyoung Lee, Dong-Ha Kim, Joon-Young Kang, et al. 2019. Heterogeneous metal oxide–graphene thorn-bush single fiber as a freestanding chemiresistor. ACS Applied Materials & Interfaces 11(10): 10208–10217. American Chemical Society. doi:10.1021/acsami.8b22015.

Jeon, Minsu, Bongsik Choi, Jinsu Yoon, Dong Myong Kim, Dae Hwan Kim, Inkyu Park and Sung-Jin Choi. 2017. Enhanced sensing of gas molecules by a 99.9% semiconducting carbon nanotube-based field-effect transistor sensor. Applied Physics Letters 111(2): 22102. American Institute of Physics. doi:10.1063/1.4991970.

Jo, Young-Moo, Tae-Hyung Kim, Chul-Soon Lee, Kyeorei Lim, Chan Woong Na, Faissal Abdel-Hady, Abdulaziz A. Wazzan, et al. 2018. Metal–organic framework-derived hollow hierarchical Co_3O_4 nanocages with tunable size and morphology: ultrasensitive and highly selective detection of methylbenzenes. ACS Applied Materials & Interfaces 10(10): 8860–8868. American Chemical Society. doi:10.1021/acsami.8b00733.

Kalai Priya, A., Anoop Sunny, B. Karthikeyan and D. Sastikumar. 2020. Optical, spectroscopic and fiber optic gas sensing of potassium doped α-Fe_2O_3 nanostructures. Optical Fiber Technology 58: 102304. doi:https://doi.org/10.1016/j.yofte.2020.102304.

Kalyakin, Anatoly S., Dmitry A. Medvedev and Alexander N. Volkov. 2021. Electrochemical sensors based on proton-conducting electrolytes for determination of concentration and diffusion coefficient of CO_2 in inert gases. Chemical Engineering Science 229: 116046. doi:https://doi.org/10.1016/j.ces.2020.116046.

Kathiravan, Deepa, Bohr-Ran Huang and Adhimoorthy Saravanan. 2017. Self-assembled hierarchical interfaces of ZnO Nanotubes/Graphene heterostructures for efficient room temperature hydrogen sensors. ACS Applied Materials & Interfaces 9(13) : 12064–12072. American Chemical Society. doi:10.1021/acsami.7b00338.

Keley, Meysam M., Fabrício F. Borghi, Regina C. Allil, Alexandre Mello and Marcelo M. Werneck. 2021. Cu$_2$-XO-functionalized plastic optical fiber for H$_2$S sensing. Optical Fiber Technology 62: 102469. doi:https://doi.org/10.1016/j.yofte.2021.102469.

Khan, Muhammad Waqas, M. Munir Sadiq, Karuppasamy Gopalsamy, Kai Xu, Azmira Jannat, Bao Yue Zhang, Md Mohiuddin, et al. 2022. Hetero-metallic metal-organic frameworks for room-temperature NO$_2$ sensing. Journal of Colloid and Interface Science 610: 304–312. doi:https://doi.org/10.1016/j.jcis.2021.11.177.

Khudadad, Ameer I., Ali A. Yousif and Husam R. Abed. 2021. Effect of heat treatment on WO$_3$ nanostructures based NO$_2$ gas sensor low-cost device. Materials Chemistry and Physics 269: 124731. doi:https://doi.org/10.1016/j.matchemphys.2021.124731.

Kim, Jin-Young, Jae-Hyoung Lee, Jae-Hun Kim, Ali Mirzaei, Hyoun Woo Kim and Sang Sub Kim. 2019. Realization of H$_2$S sensing by Pd-functionalized networked CuO nanowires in self-heating mode. Sensors and Actuators B: Chemical 299: 126965. doi:https://doi.org/10.1016/j.snb.2019.126965.

Kitture, Rohini, Dnyandeo Pawar, Ch.N. Rao, Ravi Kant Choubey and S.N. Kale. 2017. Nanocomposite modified optical fiber: a room temperature, selective H$_2$S gas sensor: studies using ZnO-PMMA. Journal of Alloys and Compounds 695: 2091–2096. doi:https://doi.org/10.1016/j.jallcom.2016.11.048.

Kösemen, Arif, Sadullah Öztürk, Zafer Şen, Zühal Alpaslan Kösemen, Mika Harbeck and Zafer Ziya Öztürk. 2017. Volatile organic compounds and dimethyl methyl phosphonate (DMMP) sensing properties of the metal oxide functionalized QCM transducers at room temperature. Journal of The Electrochemical Society 164(13): B657–664. The Electrochemical Society. doi:10.1149/2.1251713jes.

Kumar, Manjeet, Vishwa Bhatt and Ju-Hyung Yun. 2020. Hierarchical 3D micro flower-like Co$_3$O$_4$ structures for NO$_2$ detection at room temperature. Physics Letters A 384(19): 126477. doi:https://doi.org/10.1016/j.physleta.2020.126477.

Li, Yang, Huitao Ban and Mujie Yang. 2016. Highly sensitive NH$_3$ gas sensors based on novel polypyrrole-coated SnO$_2$ nanosheet nanocomposites. Sensors and Actuators B: Chemical 224: 449–457. doi:https://doi.org/10.1016/j.snb.2015.10.078.

Li, Zhijie, Junqiang Wang, Ningning Wang, Shengnan Yan, Wei Liu, Yong Qing Fu and Zhiguo Wang. 2017. Hydrothermal synthesis of hierarchically flower-like CuO nanostructures with porous nanosheets for excellent H$_2$S sensing. Journal of Alloys and Compounds 725: 1136–1143. doi:https://doi.org/10.1016/j.jallcom.2017.07.218.

Li, Qun, Jiabin Wu, Liang Huang, Junfeng Gao, Haowen Zhou, Yijie Shi, Qinhe Pan, et al. 2018. Sulfur dioxide gas-sensitive materials based on zeolitic imidazolate framework-derived carbon nanotubes. Journal of Materials Chemistry A 6(25): 12115–12124. The Royal Society of Chemistry. doi:10.1039/C8TA02036A.

Li, Weiwei, Jiahui Guo, Li Cai, Wenzhi Qi, Yilin Sun, Jian-Long Xu, Mengxing Sun, et al. 2019. UV light irradiation enhanced gas sensor selectivity of NO$_2$ and SO$_2$ using rGO functionalized with hollow SnO$_2$

nanofibers. Sensors and Actuators B: Chemical 290: 443–452. doi:https://doi. org/10.1016/j.snb.2019.03.133.

Li, Huijun, Ning Zhang, Xiaolei Zhao, Zhouqing Xu, Zhanying Zhang and Yan Wang. 2020a. Modulation of TEA and methanol gas sensing by ion-exchange based on a sacrificial template 3D diamond-shaped MOF. Sensors and Actuators B: Chemical 315: 128136. doi:https://doi.org/10.1016/j.snb.2020. 128136.

Li, Sihan, Lili Xie, Meng He, Xiaobing Hu, Guifang Luo, Cheng Chen and Zhigang Zhu. 2020b. Metal-Organic Frameworks-Derived Bamboo-like CuO/In_2O_3 heterostructure for high-performance H_2S gas sensor with low operating temperature. Sensors and Actuators B: Chemical 310: 127828. doi:https://doi.org/10.1016/j.snb.2020.127828.

Li, Yuehua, Shaoxian Li, Xu Li, Weiwei Meng, Lei Dai and Ling Wang. 2021. Electrochemical exsolution of Ag nanoparticles from $AgNbO_3$ sensing electrode for enhancing the performance of mixed potential type NH_3 sensors. Sensors and Actuators B: Chemical 344: 130296. doi:https://doi. org/10.1016/j.snb.2021.130296.

Liang, Jiran, Junfeng Liu, Wenjiao Li and Ming Hu. 2016. Preparation and room temperature methane sensing properties of platinum-decorated vanadium oxide films. Materials Research Bulletin 84: 332–339. doi:https://doi.org/10. 1016/j.materresbull.2016.08.024.

Liang, Jiran, Ran Yang, Kuilong Zhu and Ming Hu. 2018. Room temperature acetone-sensing properties of branch-like VO_2 (B)@ZnO hierarchical hetero-nanostructures. Journal of Materials Science: Materials in Electronics 29(5): 3780–89. doi:10.1007/s10854-017-8313-4.

Liang, Yue, Cheng Wu, Shutong Jiang, Yong Jie Li, Dui Wu, Mei Li, Peng Cheng, et al. 2021. Field comparison of electrochemical gas sensor data correction algorithms for ambient air measurements. Sensors and Actuators B: Chemical 327: 128897. doi:https://doi.org/10.1016/j.snb.2020.128897.

Liang, Qihua, Yunjia Guo, Hui Chen, MeiHong Fan, Ni Bai, Xiaoxin Zou and Guo-Dong Li. 2022. Energy level regulation to optimize hydrogen sensing performance of porous bimetallic gallium-indium oxide with ultrathin pore walls. Sensors and Actuators B: Chemical 350: 130864. doi:https://doi. org/10.1016/j.snb.2021.130864.

Liu, Bohao, Xueyan Liu, Zhen Yuan, Yadong Jiang, Yuanjie Su, Jinyi Ma and Huiling Tai. 2019a. A flexible NO_2 gas sensor based on polypyrrole/ nitrogen-doped multiwall carbon nanotube operating at room temperature. Sensors and Actuators B: Chemical 295: 86–92. doi:https://doi.org/10.1016/j. snb.2019.05.065.

Liu, Tong, Luyao Li, Xinyu Yang, Xishuang Liang, Fengmin Liu, Fangmeng Liu, Chuan Zhang, et al. 2019b. Mixed potential type acetone sensor based on $Ce_{0.8}Gd_{0.2}O_{1.95}$ and $Bi_{0.5}La_{0.5}FeO_3$ sensing electrode used for the detection of diabetic ketosis. Sensors and Actuators B: Chemical 296: 126688. doi:https:// doi.org/10.1016/j.snb.2019.126688.

Liu, Hai, Wen Zhang, Haoran Wang, Cancan Chen and Shoufeng Tang. 2020. Design of methane sensor based on slow light effect in hollow core photonic crystal fiber. Sensors and Actuators A: Physical 303: 111791. doi:https://doi. org/10.1016/j.sna.2019.111791.

Liu, Xiaojing, Xinping Duan, Chong Zhang, Peiyu Hou and Xijin Xu. 2022. Improvement toluene detection of gas sensors based on flower-like porous indium oxide nanosheets. Journal of Alloys and Compounds 897: 163222. doi:https://doi.org/10.1016/j.jallcom.2021.163222.

Lopez, Juan D., Meysam Keley, Alex Dante and Marcelo M. Werneck. 2021. Optical fiber sensor coated with copper and iron oxide nanoparticles for hydrogen sulfide sensing. Optical Fiber Technology 67: 102731. doi:https://doi.org/10.1016/j.yofte.2021.102731.

Łukowiec, Dariusz, Jerzy Kubacki, Piotr Kałużyński, Marcin Procek, Stanisław Wacławek and Adrian Radoń. 2022. Formation and role in gas sensing properties of spherical and hollow silver nanoparticles deposited on the surface of electrochemically exfoliated graphite. Applied Surface Science 580: 152316. doi:https://doi.org/10.1016/j.apsusc.2021.152316.

Lundström, Ingemar, Hans Sundgren, Fredrik Winquist, Mats Eriksson, Christina Krantz-Rülcker and Anita Lloyd-Spetz. 2007. Twenty-five years of field effect gas sensor research in linköping. Sensors and Actuators B: Chemical 121(1): 247–262. doi:https://doi.org/10.1016/j.snb.2006.09.046.

Ma, Jiangwei, Huiqing Fan, Zhexin Li, Yuxin Jia, Arun Kumar Yadav, Guangzhi Dong, Weijia Wang, et al. 2021. Multi-walled carbon nanotubes/polyaniline on the ethylenediamine modified polyethylene terephthalate fibers for a flexible room temperature ammonia gas sensor with high responses. Sensors and Actuators B: Chemical 334: 129677. doi:https://doi.org/10.1016/j.snb.2021.129677.

Meng, Leixin, Qi Xu, Zhe Sun, Gaoda Li, Suo Bai, Zenghua Wang and Yong Qin. 2018. Enhancing the performance of room temperature ZnO microwire gas sensor through a combined technology of surface etching and UV illumination. Materials Letters 212: 296–298. doi:https://doi.org/10.1016/j.matlet.2017.10.102.

Mohan Reddy, A. Jagan, M.S. Surendra Babu and P. Nagaraju. 2022. ZnNi(NA) (NA= nicotinic acid) bimetallic mesoporous MOFs as a sensing platform for ethanol, formaldehyde and ammonia at room temperature. Solid State Sciences 125: 106819. doi:https://doi.org/10.1016/j.solidstatesciences.2022.106819.

Mousavi, Saeb, Kyungnam Kang, Jaeho Park and Inkyu Park. 2016. A room temperature hydrogen sulfide gas sensor based on electrospun polyaniline–polyethylene oxide nanofibers directly written on flexible substrates. RSC Advances 6(106): 104131–104138. The Royal Society of Chemistry. doi:10.1039/C6RA20710C.

Muthukrishnan, K., M. Vanaraja, S. Boomadevi, R.K. Karn, V. Singh, P.K. Singh and K. Pandiyan. 2016. Studies on acetone sensing characteristics of ZnO thin film prepared by Sol–Gel dip coating. Journal of Alloys and Compounds 673: 138–143. doi:https://doi.org/10.1016/j.jallcom.2016.02.222.

Naik, Shreyanka Shankar, Seung Jun Lee, Jayaraman Theerthagiri, Yiseul Yu and Myong Yong Choi. 2021. Rapid and highly selective electrochemical sensor based on ZnS/Au-Decorated f-Multi-Walled carbon nanotube nanocomposites produced via pulsed laser technique for detection of toxic nitro compounds. Journal of Hazardous Materials 418: 126269. doi:https://doi.org/10.1016/j.jhazmat.2021.126269.

Nasresfahani, Sh., S. Javanmardi, M.H. Sheikhi and M. Khalilakbar. 2021. Enhanced relatively low-temperature carbon monoxide sensing properties of cupric oxide/porous graphitic carbon nitride p–n heterojunction. Sensors and Actuators A: Physical 331: 113004. doi:https://doi.org/10.1016/j sna.2021. 113004.

Nasri, Abdelghaffar, Mathieu Pétrissans, Vanessa Fierro and Alain Celzard. 2021. Gas Sensing based on organic composite materials: review of sensor types, progresses and challenges. Materials Science in Semiconductor Processing 128: 105744. doi:https://doi.org/10.1016/j.mssp.2021.105744.

Nguyen, Duy-Khoi, Jae-Hyoung Lee, Tan Le-Hoang Doan, Thanh-Binh Nguyen, Sungkyun Park, Sang Sub Kim and Bach Thang Phan. 2020. H_2 gas sensing of co-incorporated metal-organic frameworks. Applied Surface Science 523: 146487. doi:https://doi.org/10.1016/j.apsusc.2020.146487.

Öztürk, Sadullah, Arif Kösemen, Zühal Alpaslan Kösemen, Necmettin Kılınç, Zafer Ziya Öztürk and Michele Penza. 2016. Electrochemically growth of Pd doped ZnO nanorods on QCM for room temperature VOC sensors. Sensors and Actuators B: Chemical 222: 280–289. doi:https://doi.org/10.1016/j.snb. 2015.08.083.

Park, Sunghoon, Hyejoon Kheel, Gun-Joo Sun, Hyoun Woo Kim, Taegyung Ko and Chongmu Lee. 2016. Room-temperature hydrogen gas sensing properties of the networked Cr_2O_3-functionalized Nb_2O_5 nanostructured sensor. Metals and Materials International 22(4): 730–736. doi:10.1007/s12540-016-6028-3.

Patel, Lisa, Elizabeth Friedman, Stephanie Alexandra Johannes, Stephanie Sophie Lee, Haley Grace O'Brien and Sarah E Schear. 2021. Air pollution as a social and structural determinant of health. The Journal of Climate Change and Health 3: 100035. doi:https://doi.org/10.1016/j.joclim.2021.100035.

Paul, Jiss and Jacob Philip. 2022. Development of an ammonia gas sensor employing an inter-digital capacitive structure coated with $Mn_{0.5}Zn_{0.5}Fe_2O_4$ nano-composite. Materials Today: Proceedings 49: 1331–1336. doi:https://doi.org/10.1016/j.matpr.2021.06.407.

Pawar, Dnyandeo, Rohini Kitture and S.N. Kale. 2017. ZnO coated fabry-perot interferometric optical fiber for detection of gasoline blend vapors: refractive index and fringe visibility manipulation studies. Optics & Laser Technology 89: 46–53. doi:https://doi.org/10.1016/j.optlastec.2016.09.038.

Pawar, Dnyandeo, B.V. Bhaskara Rao and S.N. Kale. 2018. Fe_3O_4-decorated graphene assembled porous carbon nanocomposite for ammonia sensing: study using an optical fiber fabry–perot interferometer. Analyst 143(8): 1890–1898. The Royal Society of Chemistry. doi:10.1039/C7AN01891F.

Peng, Fang, Shaojie Wang, Weiwei Yu, Tiantian Huang, Yan Sun, Chuanwei Cheng, Xin Chen, et al. 2020. Ultrasensitive ppb-level H_2S gas sensor at room temperature based on WO_3/rGO hybrids. Journal of Materials Science: Materials in Electronics 31(6): 5008–16. doi:10.1007/s10854-020-03067-6.

Ponhan, Witawat, Surachet Phadungdhitidhada and Supab Choopun. 2017. Fabrication of ethanol sensors based on ZnO thin film field-effect transistor prepared by thermal evaporation deposition. Materials Today: Proceedings 4 (5, Part 2): 6342–6348. doi:https://doi.org/10.1016/j.matpr.2017.06.137.

Qi, Pengjia, Ziying Wang, Rui Wang, Yinan Xu and Tong Zhang. 2016. Studies on QCM-type NO_2 gas sensor based on graphene composites at room temperature. Chemical Research in Chinese Universities 32(6): 924–928. doi:10.1007/s40242-016-6129-z.

Qin, Xiang, Wenlin Feng, Xiaozhan Yang, Jianwei Wei and Guojia Huang. 2018. Molybdenum sulfide/citric acid composite membrane-coated long period fiber grating sensor for measuring trace hydrogen sulfide gas. Sensors and Actuators B: Chemical 272: 60–68. doi:https://doi.org/10.1016/j.snb.2018.05.152.

Ramu, A.G., A. Umar, S. Gopi, H. Algadi, H. Albargi, A.A. Ibrahim, M.A. Alsaiari, et al. 2021. Tetracyanonickelate (II)/KOH/reduced graphene oxide fabricated carbon felt for mediated electron transfer type electrochemical sensor for efficient detection of N_2O gas at room temperature. Environmental Research 201: 111591. doi:https://doi.org/10.1016/j.envres.2021.111591.

Reddeppa, M., B.-G. Park, M.-D. Kim, K.R. Peta, N.D. Chinh, D. Kim, S.-G. Kim, et al. 2018. H_2, H_2S gas sensing properties of rGO/GaN nanorods at room temperature: effect of UV illumination. Sensors and Actuators B: Chemical 264: 353–362. doi:https://doi.org/10.1016/j.snb.2018.03.018.

Reddeppa, M., N.T. KimPhung, G. Murali, K.S. Pasupuleti, B.-G. Park, Insik In and M.-D. Kim. 2021. Interaction activated interfacial charge transfer in 2D G-C3N4/GaN nanorods heterostructure for self-powered UV photodetector and room temperature NO_2 gas sensor at ppb level. Sensors and Actuators B: Chemical 329: 129175. doi:https://doi.org/10.1016/j.snb.2020.129175.

Renganathan, B., Subha Krishna Rao, A.R. Ganesan and A. Deepak. 2021. High proficient sensing response in clad modified ceria doped tin oxide fiber optic toxic gas sensor application. Sensors and Actuators A: Physical 332: 113114. doi:https://doi.org/10.1016/j.sna.2021.113114.

Rithesh Raj, D., S. Prasanth, T.V. Vineeshkumar and C. Sudarsanakumar. 2015. Ammonia sensing properties of tapered plastic optical fiber coated with silver nanoparticles/PVP/PVA hybrid. Optics Communications 340: 86–92. doi:https://doi.org/10.1016/j.optcom.2014.11.092.

Sapsanis, Christos, Hesham Omran, Valeriya Chernikova, Osama Shekhah, Youssef Belmabkhout, Ulrich Buttner, Mohamed Eddaoudi, et al. 2015. Insights on capacitive interdigitated electrodes coated with MOF thin films: humidity and VOCs sensing as a case study. Sensors. doi:10.3390/s150818153.

Saravia, Lucas P.H., A. Sukeri, Josue M. Gonçalves, Juan S. Aguirre-Araque, Bruno B.N.S. Brandão, Tiago A. Matias, Marcelo Nakamura. 2016. CoTRP/graphene oxide composite as efficient electrode material for dissolved oxygen sensors. Electrochimica Acta 222: 1682–1690. doi:https://doi.org/10.1016/j.electacta.2016.11.159.

SebtAhmadi, S.S., B. Raissi, M. Sahba Yaghmaee, R. Riahifar and S. Rahimisheikh. 2021. Effect of electrode pores on the performance of CO electrochemical gas sensor, experimental and modeling. Electrochimica Acta 389: 138611. doi:https://doi.org/10.1016/j.electacta.2021.138611.

Sen, Sovandeb and Susmita Kundu. 2021. Reduced graphene oxide (rGO) decorated $ZnO-SnO_2$: a ternary nanocomposite towards improved low concentration VOC sensing performance. Journal of Alloys and Compounds 881: 160406. doi:https://doi.org/10.1016/j.jallcom.2021.160406.

Shankar, Prabakaran and John Bosco Balaguru Rayappan. 2017. Room temperature ethanol sensing properties of ZnO nanorods prepared using an electrospinning technique. Journal of Materials Chemistry C 5(41): 10869–10880. The Royal Society of Chemistry. doi:10.1039/C7TC03771F.

Shin, Wonjun, Gyuweon Jung, Seongbin Hong, Yujeong Jeong, Jinwoo Park, Donghee Kim, Byung-Gook Park, et al. 2022. Optimization of channel structure and bias condition for signal-to-noise ratio improvement in Si-based FET-type gas sensor with horizontal floating-gate. Sensors and Actuators B: Chemical 357: 131398. doi:https://doi.org/10.1016/j.snb.2022.131398.

Shojaee, M., Sh. Nasresfahani and M.H. Sheikhi. 2018. Hydrothermally synthesized pd-loaded SnO_2/Partially reduced graphene oxide nanocomposite for effective detection of carbon monoxide at room temperature. Sensors and Actuators B: Chemical 254: 457–467. doi:https://doi.org/10.1016/j.snb.2017.07.083.

Shukla, Prashant, Pooja Saxena, Devinder Madhwal, Nitin Bhardwaj and V.K. Jain. 2021. Battery-operated resistive sensor based on electrochemically exfoliated pencil graphite core for room temperature detection of LPG. Sensors and Actuators B: Chemical 343: 130133. doi:https://doi.org/10.1016/j.snb.2021.130133.

Silambarasan, P. and I.S. Moon. 2021. Real-time monitoring of chlorobenzene gas using an electrochemical gas sensor during mediated electrochemical degradation at room temperature. Journal of Electroanalytical Chemistry 894: 115372. doi:https://doi.org/10.1016/j.jelechem.2021.115372.

Silva, Luís F. da, J.-C. M'Peko, Ariadne C. Catto, Sandrine Bernardini, Valmor R. Mastelaro, Khalifa Aguir, Caue Ribeiro, et al. 2017. UV-enhanced ozone gas sensing response of ZnO-SnO_2 heterojunctions at room temperature. Sensors and Actuators B: Chemical 240: 573–579. doi:https://doi.org/10.1016/j.snb.2016.08.158.

Singh, Pratibha, Chandra Shekhar Kushwaha, Vinay Kumar Singh, G.C. Dubey and Saroj Kr. Shukla. 2021. Chemiresistive sensing of volatile ammonia over zinc oxide encapsulated polypyrrole based nanocomposite. Sensors and Actuators B: Chemical 342: 130042. doi:https://doi.org/10.1016/j.snb.2021.130042.

Song, Zhilong, Zeru Wei, Baocun Wang, Zhen Luo, Songman Xu, Wenkai Zhang, Haoxiong Yu, et al. 2016. Sensitive room-temperature H_2S gas sensors employing SnO_2 quantum wire/reduced graphene oxide nanocomposites. Chemistry of Materials 28(4): 1205–1212. American Chemical Society. doi:10.1021/acs.chemmater.5b04850.

Sridevi., S., K.S. Vasu, Navakanta Bhat, S. Asokan and A.K. Sood. 2016. Ultra sensitive NO_2 gas detection using the reduced graphene oxide coated etched fiber bragg gratings. Sensors and Actuators B: Chemical 223: 481–486. doi:https://doi.org/10.1016/j.snb.2015.09.128.

Subha, P.P., K. Hasna and M.K. Jayaraj. 2017. Surface modification of TiO_2 nanorod arrays by Ag nanoparticles and its enhanced room temperature ethanol sensing properties. Materials Research Express 4(10): 105037. IOP Publishing. doi:10.1088/2053-1591/aa91ee.

Subin David, S.P., S. Veeralakshmi, J. Sandhya, S. Nehru and S. Kalaiselvam. 2020. Room temperature operatable high sensitive toluene gas sensor using chemiresistive Ag/Bi$_2$O$_3$ nanocomposite. Sensors and Actuators B: Chemical 320: 128410. doi:https://doi.org/10.1016/j.snb.2020.128410.

Sun, C., P.R. Ohodnicki and Y. Yu. 2017. Double-layer zeolite nano-blocks and palladium-based nanocomposite fiber optic sensors for selective hydrogen sensing at room temperature. IEEE Sensors Letters 1(5): 1–4. doi:10.1109/LSENS.2017.2754142.

Sun, Yongjiao, Zhenting Zhao, Koichi Suematsu, Pengwei Li, Wendong Zhang and Jie Hu. 2022. Moisture-resisting acetone sensor based on MOF-derived ZnO-NiO nanocomposites. Materials Research Bulletin 146: 111607. doi:https://doi.org/10.1016/j.materresbull.2021.111607.

Surya, Sandeep G., Sreenu Bhanoth, Sanjit M. Majhi, Yogeshwar D. More, V. Mani Teja and Karumbaiah N. Chappanda. 2019. A silver nanoparticle-anchored UiO-66(Zr) metal–organic framework (MOF)-based capacitive H$_2$S gas sensor. CrystEngComm 21(47): 7303–7312. The Royal Society of Chemistry. doi:10.1039/C9CE01323G.

Tabassum, S., R. Kumar and L. Dong. 2017. Nanopatterned optical fiber tip for guided mode resonance and application to gas sensing. IEEE Sensors Journal 17(22): 7262–7272. doi:10.1109/JSEN.2017.2748593.

Tabata, Hiroshi, Yuta Sato, Kouhei Oi, Osamu Kubo and Mitsuhiro Katayama. 2018. Bias- and gate-tunable gas sensor response originating from modulation in the schottky barrier height of a graphene/MoS$_2$ van Der Waals heterojunction. ACS Applied Materials & Interfaces 10(44): 38387–38393. American Chemical Society. doi:10.1021/acsami.8b14667.

Tang, Shi, Wenying Ma, Guangzhong Xie, Yuanjie Su and Yadong Jiang. 2016. Acetylcholinesterase-reduced graphene oxide hybrid films for organophosphorus neurotoxin sensing via quartz crystal microbalance. Chemical Physics Letters 660: 199–204. doi:https://doi.org/10.1016/j.cplett.2016.08.025.

Tharsika, T., M. Thanihaichelvan, A.S.M.A. Haseeb and S.A. Akbar. 2019. Highly sensitive and selective ethanol sensor based on ZnO nanorod on SnO$_2$ thin film fabricated by spray pyrolysis. Frontiers in Materials. https://www.frontiersin.org/article/10.3389/fmats.2019.00122.

Toan, Nguyen Van, Chu Manh Hung, Nguyen Van Duy, Nguyen Duc Hoa, Dang Thi Thanh Le and Nguyen Van Hieu. 2017. Bilayer SnO$_2$–WO$_3$ nanofilms for enhanced NH$_3$ gas sensing performance. Materials Science and Engineering: B 224: 163–70. doi:https://doi.org/10.1016/j.mseb.2017.08.004.

Tolentino, Maria Ai Kristine P., Dharmatov Rahula B. Albano and Fortunato B. Sevilla. 2018. Piezoelectric sensor for ethylene based on Silver(I)/polymer composite. Sensors and Actuators B: Chemical 254: 299–306. doi:https://doi.org/10.1016/j.snb.2017.07.015.

Tshabalala, Z.P., K. Shingange, B.P. Dhonge, O.M. Ntwaeaborwa, G.H. Mhlongo and D.E. Motaung. 2017. Fabrication of ultra-high sensitive and selective CH$_4$ room temperature gas sensing of TiO$_2$ nanorods: detailed study on the annealing temperature. Sensors and Actuators B: Chemical 238: 402–419. doi:https://doi.org/10.1016/j.snb.2016.07.023.

Wang, Ziying, Tong Zhang, Tianyi Han, Teng Fei, Sen Liu and Geyu Lu. 2018. Oxygen vacancy engineering for enhanced sensing performances: a case of

SnO$_2$ nanoparticles-reduced graphene oxide hybrids for ultrasensitive ppb-level room-temperature NO$_2$ sensing. Sensors and Actuators B: Chemical 266: 812–822. doi:https://doi.org/10.1016/j.snb.2018.03.169.

Wang, Hui, Yue Wang, Xiaopeng Hou and Benhai Xiong. 2020. Bioelectronic nose based on single-stranded DNA and single-walled carbon nanotube to identify a major plant volatile organic compound (p-Ethylphenol) released by phytophthora cactorum infected strawberries. Nanomaterials 10(3): 479. https://doi.org/10.3390/nano10030479.

Wang, Luyu and Jia Song. 2020. Enhanced NH$_3$ sensing properties of carboxyl functionalized carbon nanocoil. Materials Research Express 7(7): 75014. IOP Publishing. doi:10.1088/2053-1591/aba806.

Wang, Wei, Yuanyi Zhang, Jinniu Zhang, Gang Li, Deying Leng, Ying Gao, Jianzhi Gao, et al. 2021a. Metal–organic framework-derived Cu$_2$O–CuO octahedrons for sensitive and selective detection of ppb-level NO$_2$ at room temperature. Sensors and Actuators B: Chemical 328: 129045. doi:https://doi.org/10.1016/j.snb.2020.129045.

Wang, Jing, Chenyu Hu, Yi Xia and Bo Zhang. 2021b. Mesoporous ZnO nanosheets with rich surface oxygen vacancies for UV-activated methane gas sensing at room temperature. Sensors and Actuators B: Chemical 333: 129547. doi:https://doi.org/10.1016/j.snb.2021.129547.

Wang, Jing, Yuyan Yang and Yi Xia. 2022. Mesoporous MXene/ZnO nanorod hybrids of high surface area for UV-activated NO$_2$ gas sensing in ppb-level. Sensors and Actuators B: Chemical 353: 131087. doi:https://doi.org/10.1016/j.snb.2021.131087.

Wei, Yuling, Changlong Chen, Guangzheng Yuan and Shuai Gao. 2016. SnO$_2$ nanocrystals with abundant oxygen vacancies: preparation and room temperature NO$_2$ sensing. Journal of Alloys and Compounds 681: 43–49. doi:https://doi.org/10.1016/j.jallcom.2016.04.220.

Wei, Xiongbang, Xiaohui Yang, Tao Wu, Shuanghong Wu, Weizhi Li, Xiaohui Wang and Zhi Chen. 2017. A novel hydrogen-sensitive sensor based on Pd Nanorings/TNTs composite structure. International Journal of Hydrogen Energy 42(38): 24580–24586. doi:https://doi.org/10.1016/j.ijhydene.2017.07.167.

Wei, Shuangjiao, Zhenglan Bian, Xingquan Wang, Guilin Zhang, Liang Xue, Anduo Hu and Fenghong Chu. 2022. Performance of transmitted optical fiber hydrogen sensing system based on orthogonal demodulation method. Optics Communications 502: 127401. doi:https://doi.org/10.1016/j.optcom.2021.127401.

Wu, Baofeng, Linlin Wang, Hongyuan Wu, Kan Kan, Guo Zhang, Yu Xie, Ye Tian, et al. 2016a. Templated synthesis of 3D hierarchical porous Co$_3$O$_4$ materials and their NH$_3$ sensor at room temperature. Microporous and Mesoporous Materials 225: 154–163. doi:https://doi.org/10.1016/j.micromeso.2015.12.019.

Wu, Chao-Wei, Chien-Chung Wu and Chia-Chin Chiang. 2016b. A ZnO nanoparticle-coated long period fiber grating as a carbon dioxide gas sensor. Inventions 1(4): 21. https://doi.org/10.3390/inventions1040021

Wu, Chenggen, Lei Han, Jichao Zhang, Yongqing Wang, Rui Wang and LiJun Chen. 2021. Capacitive ammonia sensor based on graphene oxide/polyaniline nanocomposites. Advanced Materials Technologies 7(7): 2101247. John Wiley & Sons, Ltd. doi:https://doi.org/10.1002/admt.202101247.

Xia, Yi, Jing Wang, Lei Xu, Xian Li and Shaojun Huang. 2020. A room-temperature methane sensor based on Pd-decorated ZnO/rGO hybrids enhanced by visible light photocatalysis. Sensors and Actuators B: Chemical 304: 127334. doi.org/10.1016/j.snb.2019.127334.

Xu, Shuang, Jun Gao, Linlin Wang, Kan Kan, Yu Xie, Peikang Shen, Li Li, et al. 2015. Role of the heterojunctions in In_2O_3-composite SnO_2 nanorod sensors and their remarkable gas-sensing performance for NOx at room temperature. Nanoscale 7(35): 14643–14651. The Royal Society of Chemistry. doi:10.1039/C5NR03796D.

Yan, Fanfan, Guowen Shen, Xi Yang, Tianjiao Qi, Jie Sun, Xinghua Li and Mingtao Zhang. 2019. Low operating temperature and highly selective NH_3 chemiresistive gas sensors based on Ag_3PO_4 semiconductor. Applied Surface Science 479: 1141–1147. doi:https://doi.org/10.1016/j.apsusc.2019.02.184.

Yang, Shulin, Zhao Wang, Yongming Hu, Xiantao Luo, Jinmei Lei, Di Zhou, Linfeng Fei, et al. 2015a. Highly responsive room-temperature hydrogen sensing of α-MoO_3 nanoribbon membranes. ACS Applied Materials & Interfaces 7(17): 9247–9253. American Chemical Society. doi:10.1021/acsami.5b01858.

Yang, Jianchun, Lang Zhou, Jing Huang, Chuanyi Tao, Xueming Li and Weimin Chen. 2015b. Sensitivity enhancing of transition mode long-period fiber grating as methane sensor using high refractive index polycarbonate/cryptophane A overlay deposition. Sensors and Actuators B: Chemical 207: 477–480. doi:https://doi.org/10.1016/j.snb.2014.10.013.

Yang, Jianchun, Lang Zhou, Xin Che, Jing Huang, Xueming Li and Weimin Chen. 2016. Photonic crystal fiber methane sensor based on modal interference with an ultraviolet curable fluoro-siloxane nano-film incorporating cryptophane A. Sensors and Actuators B: Chemical 235: 717–722. doi:https://doi.org/10.1016/j.snb.2016.05.125.

Yang, Chia Ming, Tsung Cheng Chen, Yu Cheng Yang, M. Meyyappan and Chao Sung Lai. 2017. Enhanced acetone sensing properties of monolayer graphene at room temperature by electrode spacing effect and UV illumination. Sensors and Actuators, B: Chemical 253: 77–84. Elsevier B.V. doi:10.1016/j.snb.2017.06.116.

Yang, Xueli, Hao Li, Tai Li, Zezheng Li, Weifeng Wu, Chaoge Zhou, Peng Sun, et al. 2019. "Highly Efficient Ethanol Gas Sensor Based on Hierarchical SnO_2/Zn_2SnO_4 porous spheres. Sensors and Actuators B: Chemical 282: 339–346. doi:https://doi.org/10.1016/j.snb.2018.11.070.

Yang, Zhimin, Yaqing Zhang, Shang Gao, Liang Zhao, Teng Fei, Sen Liu and Tong Zhang. 2021. Hydrogen bonds-induced room-temperature detection of DMMP based on polypyrrole-reduced graphene oxide hybrids. Sensors and Actuators B: Chemical 346: 130518. doi:https://doi.org/10.1016/j.snb.2021.130518.

Yang, Zhimin, Yaqing Zhang, Liang Zhao, Teng Fei, Sen Liu and Tong Zhang. 2022. The synergistic effects of oxygen vacancy engineering and surface gold decoration on commercial SnO_2 for ppb-level DMMP sensing. Journal of Colloid and Interface Science 608: 2703–2717. doi:https://doi.org/10.1016/j.jcis.2021.10.192.

Yao, Ming-Shui, Wen-Hua Li and Gang Xu. 2021. Metal–organic frameworks and their derivatives for electrically-transduced gas sensors. Coordination Chemistry Reviews 426: 213479. doi:https://doi.org/10.1016/j.ccr.2020.213479.

Yassine, O., O. Shekhah, A.H. Assen, Y. Belmabkhout, K.N. Salama and M. Eddaoudi. 2016. H_2S sensors: fumarate-based Fcu-MOF thin film grown on a capacitive interdigitated electrode. Angewandte Chemie International Edition 55(51): 15879–15883. John Wiley & Sons, Ltd. doi:https://doi.org/10.1002/anie.201608780.

Yu, Huimin, Aifa Sun, Yangquan Liu, Yue Zhou, Ping Fan, Jingting Luo and Aihua Zhong. 2021. Capacitive sensor based on GaN honeycomb nanonetwork for ultrafast and low temperature hydrogen gas detection. Sensors and Actuators B: Chemical 346: 130488. doi:https://doi.org/10.1016/j.snb.2021.130488.

Zhang, Ya-nan, Qilu Wu, Huijie Peng and Yong Zhao. 2016. Photonic crystal fiber modal interferometer with Pd/WO_3 coating for real-time monitoring of dissolved hydrogen concentration in transformer oil. Review of Scientific Instruments 87(12): 125002. American Institute of Physics. doi:10.1063/1.4971321.

Zhang, Jingle, Haiwei Fu, Jijun Ding, Min Zhang and Yi Zhu. 2017a. Graphene-oxide-coated interferometric optical microfiber ethanol vapor sensor. Applied Optics 56(31): 8828–8831. OSA. doi:10.1364/AO.56.008828.

Zhang, J.Y., E.J. Ding, S.C. Xu, Z.H. Li, X.X. Wang and F. Song. 2017b. Sensitization of an optical fiber methane sensor with graphene. Optical Fiber Technology 37: 26–29. doi:https://doi.org/10.1016/j.yofte.2017.06.011.

Yuan, Shuang, Xiangjie Bo and Liping Guo. 2018. In-situ growth of iron-based metal-organic framework crystal on ordered mesoporous carbon for efficient electrocatalysis of p-Nitrotoluene and Hydrazine. Analytica Chimica Acta 1024: 73–83. doi:https://doi.org/10.1016/j.aca.2018.03.064.

Yuan, Hongye, Jifang Tao, Nanxi Li, Avishek Karmakar, Chunhua Tang, Hong Cai, Stephen John Pennycook, et al. 2019. On-chip tailorability of capacitive gas sensors integrated with metal–organic framework films. Angewandte Chemie International Edition 58(40): 14089–14094. John Wiley & Sons, Ltd. doi:https://doi.org/10.1002/anie.201906222.

Zhang, Bo, Guannan Liu, Ming Cheng, Yuan Gao, Lianjing Zhao, Shan Li, Fangmeng Liu, et al. 2018. The preparation of reduced graphene oxide-encapsulated α-Fe_2O_3 hybrid and its outstanding NO_2 gas sensing properties at room temperature. Sensors and Actuators B: Chemical 261: 252–263. doi:https://doi.org/10.1016/j.snb.2018.01.143.

Zhang, Su, Lijia Zhao, Baoyu Huang and Xiaogan Li. 2020a. UV-activated formaldehyde sensing properties of hollow $TiO_2@SnO_2$ heterojunctions at room temperature. Sensors and Actuators B: Chemical 319: 128264. doi:https://doi.org/10.1016/j.snb.2020.128264.

Zhang, Yajie, Junxin Zhang, Yadong Jiang, Zaihua Duan, Bohao Liu, Qiuni Zhao, Si Wang, et al. 2020b. Ultrasensitive flexible NH_3 gas sensor based on polyaniline/$SrGe_4O_9$ nanocomposite with ppt-level detection ability at room temperature. Sensors and Actuators B: Chemical 319: 128293. doi:https://doi.org/10.1016/j.snb.2020.128293.

Zhang, Rui, Lihui Lu, Yangyang Chang and Meng Liu. 2022a. Gas sensing based on metal-organic frameworks: concepts, functions, and developments. Journal of Hazardous Materials 429: 128321. doi:https://doi.org/10.1016/j.jhazmat.2022.128321.

Zhang, Chong, Xiaoman Chen, Xiaohang Liu, Changyu Shen, Zhenlin Huang, Zhihao Wang, Tingting Lang, et al. 2022b. High sensitivity hydrogen sensor based on tilted fiber bragg grating coated with PDMS/WO_3 Film. International Journal of Hydrogen Energy 47(9): 6415–6420. doi:https://doi.org/10.1016/j.ijhydene.2021.11.238.

Zhang, Dongzhi, Zhanjia Kang, Xiaohua Liu, Jingyu Guo and Yan Yang. 2022c. Highly sensitive ammonia sensor based on PSS doped ZIF-8-derived porous carbon/polyaniline hybrid film coated on quartz crystal microbalance. Sensors and Actuators B: Chemical 357: 131419. doi:https://doi.org/10.1016/j.snb.2022.131419.

Zhang, Chong, Changyu Shen, Xiaohang Liu, Shuyi Liu, Hongchen Chen, Zhenlin Huang, Zhihao Wang, et al. 2022d. Pd/Au nanofilms based tilted fiber bragg grating hydrogen sensor. Optics Communications 502: 127424. doi:https://doi.org/10.1016/j.optcom.2021.127424.

Zhao, Jing, Mengqing Hu, Yan Liang, Qiulin Li, Xinye Zhang and Zhenyu Wang. 2020. A room temperature sub-ppm NO_2 gas sensor based on WO_3 hollow spheres. New Journal of Chemistry 44(13): 5064–5070. The Royal Society of Chemistry. doi:10.1039/C9NJ06384F.

Zhou, Jiao, Muhammad Ikram, Afrasiab Ur Rehman, Jing Wang, Yiming Zhao, Kan Kan, Weijun Zhang, et al. 2018. Highly selective detection of NH_3 and H_2S using the pristine CuO and mesoporous In_2O_3@CuO multijunctions nanofibers at room temperature. Sensors and Actuators B: Chemical 255: 1819–1830. doi:https://doi.org/10.1016/j.snb.2017.08.200.

Zhu, Dan, Yongming Fu, Weili Zang, Yayu Zhao, Lili Xing and Xinyu Xue. 2016. Room-temperature self-powered ethanol sensor based on the piezo-surface coupling effect of heterostructured α-Fe_2O_3/ZnO nanowires. Materials Letters 166: 288–291. doi:https://doi.org/10.1016/j.matlet.2015.12.106.

Zou, Yihui, Shuai Chen, Jin Sun, Jingquan Liu, Yanke Che, Xianghong Liu, Jun Zhang, et al. 2017. Highly efficient gas sensor using a hollow SnO_2 Microfiber for triethylamine detection. ACS Sensors 2(7): 897–902. American Chemical Society. doi:10.1021/acssensors.7b00276.

Chapter 6

Ion Sensor

6.1 INTRODUCTION

Heavy metal ions are toxic in nature, harmful, and non-biodegradable, and therefore, create serious impacts on the environment, human health, and aquatic life (Mishra et al. 2019). Heavy metals are highly dense elements that are extremely toxic even in small concentrations. Table 6.1 shows the serious health hazards of these metal ions and their corresponding sources. Various recognized committees have set the permissible concentration of these metal ions as tabulated in Table 6.2. They always exist in the environment due to their non-degradability. Thus, it is essential to develop a sensitive, selective, lower limit of detection (LOD), stable, and reliable ion sensor.

Several conventional techniques like inductively coupled plasma-optical emission spectrometry (ICP-OES), inductively coupled plasma-coupled to mass spectrometry (ICP-MS) or atomic absorption spectrometry (AAS) have been reported for ion sensing. They have shown good sensitivity, selectivity, and reliability. However, these techniques are time consuming, necessitate sample preparation, need a trained operator, and complex instruments, and are expensive (Xu et al. 2022a). Thus, there is a need for other methods which could overcome these drawbacks.

This chapter reviews some advanced techniques such as electrochemical, field effect transistors, fiber optics, fluorescence, and colorimetry for the effective detection of ions. In detail, the integration of materials, structures, sensing performance, and merits and demerits of each technique are summarized.

Table 6.1 The drinking water standards in WHO, main source, and major harms to humans from heavy metals in elemental states [Reproduced with permission (Jaishankar et al. 2014), Copyright 2014 SETOX & Institute of Experimental Pharmacology and Toxicology, SASc.]

Heavy metals	Maximum Content (mg/L)	Main sources	Harms to the humans
Hg	0.006	Instrument factory, salt electrolysis, precious metal smelting, cosmetics, coal burning, etc.	Great damage to brain, kidneys and the developing fetus
Pb	0.01	Various paints, gasolines, batteries, smelting, electroplating, cosmetics, hair dyes, tableware, coal, etc.	Causing harm to human brain cells, especially the fetal neural plate, mental retardation; causing dementia and brain death in the elderly
Cd	0.003	Wastewater from coal, mining, smelting, fuel, battery and chemical industries	Causing high blood pressure, renal dysfunction, cardiovascular and cerebrovascular diseases; destroying bone calcium
Cr	0.05	Inferior cosmetic raw materials, leather preparations, chrome-plated parts of metal parts, industrial pigments, tanned leather, rubber and ceramic raw materials, etc.	Numbness of limbs; ulcers, mental disorder
As	0.01	Mining, smelting of arsenic and arsenic-containing metals; production of glass, pigments, raw materials, papers using arsenic or arsenic compounds as raw materials; combustion of coal	Skin lesions; harm to nervous system; digestive and cardiovascular disorders; causing cancer of lungs, liver, bladder and skin

Table 6.2 Standard guideline values for the maximum permissible limit of heavy metals ions in drinking water recommended by the WHO, EPA, and BIS [Reproduced with permission (Nigam et al. 2021), Copyright 2021, Elsevier]

Heavy metal ions	Standard guideline values (in ppm)		
	WHO	EPA	BIS
As^{3+}	0.01	0.01	0.01
Cd^{2+}	0.003	0.005	0.003
Cr^{3+}	0.05	0.1	0.05
Cu^{2+}	2	1.3	0.05
Pb^{2+}	0.01	0.015	0.01
Hg^{2+}	0.001	0.002	0.001

6.2   ELECTROCHEMICAL-BASED ION SENSORS

There are numerous electrochemical techniques such as potentiometry, amperometry, differential pulse voltammetry (DPV), square wave voltammetry (SWV), differential pulse anodic stripping voltammetry (DPASV), square wave anodic stripping voltammetry (SWASV), cyclic voltammetry (CV), and electrochemical impedance spectroscopy (EIS). These techniques are much faster, easy to set up, cheap, and highly used due to their sensitive and selective nature. Some development of ion sensors by using an electrochemical sensor is shown in Figure 6.1. Table 6.3 shows a summary of a few electrochemical-based ion sensors.

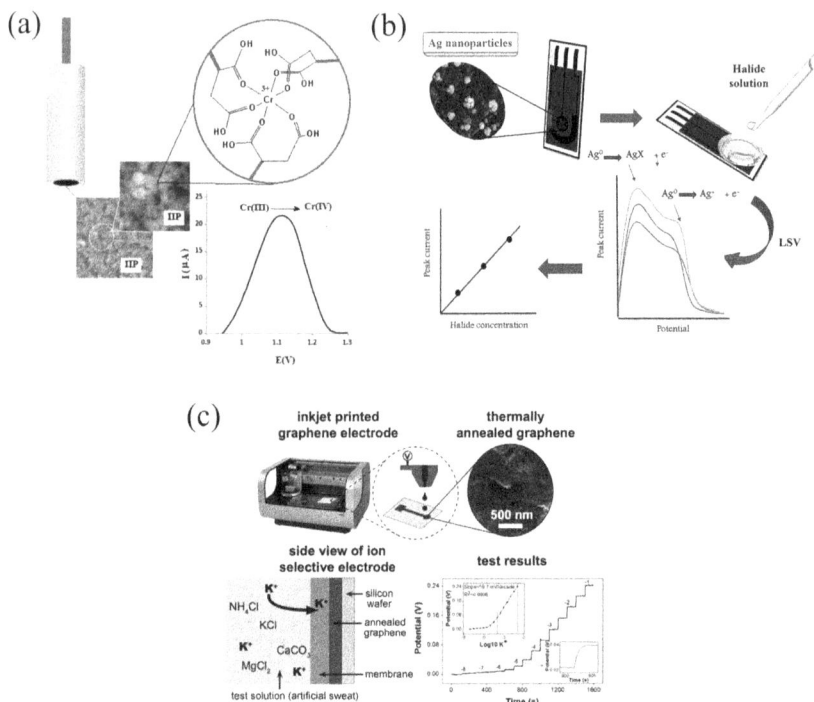

Figure 6.1   (a) Differential pulse voltammetry responses of CCE, NIP-CCE, and IIP-CCE electrodes to the stable concentration of Cr(III) at various experimental conditions, extraction condition: extraction time = 10 min, stirring rate = 500 rpm, [Cr^{3+}] = 5 µmol L^{-1}; voltametric condition: pulse time = 2 ms, modulation amplitude = 0.3 V, scan rate = 10 mV s^{-1}; analysis medium: 10 mL of acetate buffer (pH = 6, 0.1 mol L^{-1}). [Reproduced with permission (Alizadeh et al. 2017). Copyright 2017, Elsevier]. (b) Voltammetry based screen-printed electrodes to halide ions measurement based on AgNP. [Reproduced with permission (Norouzi and Parsa 2018). Copyright 2018, Elsevier]. (c) Inkjet printed graphene (IPG) for potassium ion sensing and the graph of potassium calibration plot of IPG electrodes showing a potential response curve with potassium concentration. [Reproduced with permission (He et al. 2017). Copyright 2017, American Chemical Society].

Table 6.3 A summary of a few electrochemical based ion sensor

Materials used	Technique used	Ion analyte	Detection range	Limit of detection (LOD)	References
Thiolated calix[4] arene (TC4)	DPV	Pb^{2+}; Cu^{2+}	4.83×10^{-7} M – 4.83×10^{-6} M (for Pb^{2+}), 1.57×10^{-6} M – 1.57×10^{-5} M (for Cu^{2+})	3.85×10^{-8} M (for Pb^{2+}) and 2.1×10^{-7} M (for Cu^{2+})	(Mei et al. 2021)
G4/Ag-rGO/CPE	DPV	Ba^{2+}	0.06 – 0.80 nM and 1.0 – 80 nM	0.045 nM	(Ebrahimi et al. 2022)
Ag-based electrodes	DPV	Cl^-	1 – 50 mM	20 μM	(Patella et al. 2022)
Au@GCE	DPV	NO_2^-	0.01– 3.8 mM	2.4 μM	(Shi et al. 2022)
GCE (IIP)	DPV	UO_2^{2+}	4.70×10^{-9} – 3.54×10^{-7} mol/L	1.59×10^{-9} mol/L	(Hojatpanah et al. 2022)
Sr@FeNi-S/ SWCNTs	DPV	Hg^{2+}	0.05 μM – 279 μM	0.52 nM	(Mariyappan et al. 2020)
Cr (III)-imprinted polymer (IP-CCE)	DPV	Cr^{3+}	1.00×10^{-7} – 1.00×10^{-5} mol/L	1.76×10^{-8} mol/L	(Alizadeh et al. 2017)
Magnetic silver ion imprinted polymer nanoparticles (mag-IIP-NPs)	DPV	Ag^+	4.64×10^{-10} – 1.39×10^{-6} mol/L	1.39×10^{-10} mol/L	(Ghanei-Motlagh and Taher 2017a)
Cadmium (II) ion imprinted polymer (Cd–IIP)	DPV	Cd^{2+}	4.45×10^{-9} – 3.56×10^{-7} mol/L	1.33×10^{-9} mol/L	(Ghanei-Motlagh and Taher 2017b)
Calix[4]arene-functionalized Mn_3O_4 nanoparticle	DPV	Cd^{2+}; Pb^{2+}	4.82×10^{-7} – 8.90×10^{-6} M	8.01×10^{-8} (for Cd^{2+}); 3.38×10^{-8} M (for Pb^{2+})	(Adarakatti et al. 2019)
MoS_2	DPV	Hg^{2+}	0.05 nM – 20 nM	0.029 nM	(Wang et al. 2022b)
3.3′5.5′ tetramethylbenzidine-silica nanoparticles-palladium-carbon nanoparticles (Pd/C NPs)	AMP	PO_4^{3-}	50 – 1000 μM	0.983 μM	(Altuner et al. 2022)
GC/OMC/ Ca^{2+}-ISE	AMP	Ca^{2+}	10^{-6} – 10^{-2} M	–	(Sun et al. 2022)
rGO/ZnO/Nafion	AMP	NO_2^-	20 – 520 μM	1.36 μM	(Rashed et al. 2020)
Ni/poly(4-AB)/ SDS/CPE	AMP	SO_3^{2-}	1 – 10 mM	0.063 mM	(Norouzi and Parsa 2018)

(Contd.)

Table 6.3 A summary of a few electrochemical based ion sensor (*Contd.*)

Materials used	Technique used	Ion analyte	Detection range	Limit of detection (LOD)	References
p-*tert*-butylcalix[4]arene-bis-cyrhetrenylimine (Cy_2(Calix [4])	SWV	Cu^{2+}	1.57×10^{-7} –2.83×10^{-6} M	4.72×10^{-10} M	(Pizarro et al. 2019)
GO/UiO-67@PtNPs	SWV	As^{3+}	2.7 – 40 nM	0.42 nM	(Ru et al. 2022)
Co-TMC4R-BDC/GCE	SWV	Cu^{2+}	0.05 – 12.0 μM	13 nM	(Wang et al. 2022a)
		Pb^{2+}	0.05 – 13.0 μM	11 nM	
		Cd^{2+}	0.1 – 17.0 μm	26 nM	
		Hg^{2+}	0.75 – 18.0 μM	18 nM	
Graphene quantum dots	SWV	UO_2^{2+}	8.3×10^{-9} – 7.93×10^{-8} mol/L	2.07×10^{-9} mol/L	(Guin et al. 2018)
Azo-calix[4]arene/poly(3,4-ethylenedioxy-thiophene)/poly-(styrenesulfonate) (PEDOT: PSS)	Impedance	Al^{3+}	–	–	(Echabaane et al. 2017)
PVA + NaCl + MWCNT mixture + Ag/AgCl electrode	Impedance	Ca^{2+} and Cl^-	10^{-6} – 1 M; $10^{-3.7}$ – 1 M	5 μM (for Ca^{2+});	(Xu et al. 2019)
SHCNs	Impedance	Ca^{2+}	10^{-5} – 0.05 M	–	(Zhao et al. 2019)
OMC	Impedance	Ag^+	10^{-6} – 10^{-3} M	$10^{-6.8}$ M	(Yin et al. 2020)
MWCNTs	Impedance	Pb^{2+}	$10^{-8.7}$ – $10^{-2.7}$ M	10^{-10} M	(Liu et al. 2019)
MoO_2 microspheres	Impedance	K^+	10^{-5} – 10^{-3} M	$10^{-5.5}$ M	(Zeng and Qin 2017)
MOF (Ni_3HHTP_2)	Impedance	K^+	–	5.01×10^{-7} M	(Mendecki and Mirica 2018)
Nickel modified electrode	Impedance	$H_2PO_4^-$	10^{-5} – 10^{-1} M	–	(Xu et al. 2022b)
Nano-IIP/CP	Impedance	HPO_4^{2-}	10^{-5} – 10^{-1} M	–	(Alizadeh and Atayi 2018)
PEDOT	Impedance	Na^+	0.1 – 100 mM	–	(McCaul et al. 2018)

The lead and copper ions sensing by using gold and a screen-printed carbon electrode modified with a thiolated calix[4]arene derivative was proposed (Mei et al. 2021). The DPV-based system showed that the sensor displayed a good sensitivity to lead rather than copper ions. The calixarenes (the host) were explored in this work because of their ability to interact with cationic, anionic, and unbiassed molecules and provide a stable matrix, and good identification capability. It displayed good linearity to lead ion of concentration ranging between 0.2 ppm to 1.0 ppm. However, the non-conducting property of calixarenes might limit its use in electrical devices. Another lead and cadmium ion detection using metal oxides nanoparticles was reported. The combined structure of calix[4]arene-functionalized Mn_3O_4 nanoparticle was used due to its good electrocatalytic property (Adarakatti et al. 2019), because of synergistic results, higher surface area, and many functionality units. This device showed good linearity in 100 to 1000 ppb along the LOD ~7 and 9 ppb for Pb^{2+} and Cd^{2+} ions, respectively. In the selectivity study, the decrease in peak current was observed after the addition of lead and cadmium ions, because of the development of an intermediate in copper and cadmium.

The square wave anodic stripping voltammetry technique has been successfully applied to detect copper ions using a p-tert-butylcalix[4]arene-bis-cyrhetrenylimine (Cy_2(Calix[4]) (Pizarro et al. 2019). The sensing electrode surface was prepared by the drop-coating method. The linear response to 10–180 µg/L with a LOD and LOQ of 0.03 and 0.09 µg/L was observed. It was observed that at a concentration of 10 µg/L, the sensor showed a nearby response to cadmium and lead ions. The light illumination technique was also explored for metal ion detection. In this scenario, aluminum ion sensing was performed by using an impedance spectroscopy electrochemical technique (Echabaane et al. 2017). The polymer composites like poly(3,4-ethylenedioxythiophene)/poly-(styrenesulfonate) (PEDOT:PSS) with azo-calix[4]arene were successfully investigated to detect aluminum ion under light illumination. The sensing material was coated on the indium tin oxide (ITO) electrode surface by a spin coating method. It was observed that the sensing performance was significantly improved under light illumination. A smartphone-based battery-free and wireless ion sensor was proposed. The working electrode was made up by using Ag/AgCl material while the reference electrode was fabricated using an Ag/AgCl electrode covered by PVA + NaCl + MWCNT blend (Xu et al. 2019). The sensor showed a high sensitivity, repeatability, linearity, and selectivity to Ca^{2+} and Cl^- detections.

Another Cl^- ion sensor using the silver-melamine nanowires by a conducive hollow Q-graphene (QG) skeleton was also reported. The

sensor showed a good response to Cl⁻ ions of 0.25 μm–250 mm and
LOD <0.16 μm and also a refusal to other ions. The sensor performance
to Cl⁻ ion was separately analyzed in sweat and hela cells. Overall,
the sensor showed a good sensing parameter and could be applied in
many applications. A simple and effective potentiometry with solid-state
ion-selective electrodes (ISEs) for the detection of Ca^{2+} ions based on
hollow-like carbon nanospheres (HCNs) was reported (Zhao et al. 2019).
The HCNs greatly improved the constancy and decreased the potential
drift of ISEs. This sensor showed a good sensing performance in terms
of Nernst value of around 28 mV/decade to Ca^{2+} in 10^{-5} to 0.05 M. It
was well for both parameters' detection, i.e., ion and pH. The change
of Ag NPs to Ag^+ was monitored via an ordered mesoporous carbon
material (Yin et al. 2020). This Ag^+ ISE was used to track changes of
Ag^+. The outcomes were confirmed thru the ICP-MS method. This study
reported a response to $1.0 \times 10^{-6} - 1.0 \times 10^{-3}$ M AgNO₃ with a LOD of
$10^{-6.8}$ M. This work did not report the selectivity study. Metal oxides
were also explored for the detection of ions. In short, a solid-contact
ion selective electrode used MoO_2 microspheres developed to measure
potassium ions (Zeng and Qin 2017). The MoO_2 microspheres acted as
an intermediate layer and showed a response of 55 mV/decade and LOD
of $10^{-5.5}$ M to K^+ ion. In the resistance test, the sensor performance was
not much influenced due to other parameters. The long-term stability of
over 30 days was also studied. The metal–organic frameworks (MOFs)
are very effective for ion-to-electron transducers as electrochemical
sensing (Mendecki and Mirica 2018). MOFs are highly conducting and
offer good electrocatalytic properties. To fabricate the sensing region
potentiometric electrode, the 2D MOF solution caste over the carbonic
electrode is followed by a layered ion-selective membrane. This sensor
was tested against water and results showed that the material exhibited
a good hydrophobicity which resulted in a super sensing response with
low fluctuation around 11.1 μA/h. Similarly, a flexible sensory design
for sodium ion based on PEDOT as a transducer on PMMA substrate
was demonstrated (McCaul et al. 2018). The correction results were
gathered to NaCl solution of 10^{-4} m to 0.1 m. The sensing platform
was fabricated by using 3D printing and showed a Nernstian value
57 mV/decade. However, certain issues like sweating during a workout,
pressure, tension, and risky situations need to consider.

6.3   FIELD EFFECT TRANSISTOR-BASED ION SENSORS

The high electron mobility transistor (HEMT) ion sensor based on
MoS_2 with AlGaN/GaN platform was reported (Nigam et al. 2020). The

flower-shaped MoS_2 morphology was prepared via a hydrothermal method. The flower-like morphology of MoS_2 increased the number of adsorption sites and thus, improved in response. The active edges of flower shape significantly played a key role in electrochemical reactions. The fabricated sensor exhibited linearity during 0.1 ppb-100 ppb and a sensitivity of 0.64 μA/ppb to Hg^{2+} ions. These results showed that at high concentration of mercury ion (>200 ppb), the response become almost stable and no linear change was observed. The mercapto propionic acid (MPA) and glutathione (GSH) functionalized HEMT-based cadmium ion was also fabricated (Nigam et al. 2019a). The MPA and GSH materials were applied to the gate terminal and the output changes in the drain current with respect to the ion concentration of the HEMT was monitored. The response of 0.241 μA/ppb, LOD of 0.255 ppb, and response time in order of seconds were observed. After 1 ppm to 10 ppm cadmium ion concentration, the sensor did not show a linear response. The 2D material-based ion detection has been of interest to researchers. In this domain, the MoS_2 ion sensor was fabricated for the detection of cadmium ions (Li et al. 2019). This material showed good mechanical flexibility, high elastic strain limit, low noise level, large surface-to-volume ratio, and easy preparation allowing to fabricate a high-performance flexible ion sensor. For flexibility, the flexible poly(ethylene terephthalate) (PET) substrate was used. The sensor performance was monitored in the range of 50 ng/mL to 500 μg/mL and obtained LOD of 5 ng/mL with a response time of 8 s. The stability against bending was monitored but the long-term stability was not considered. The AlGaN/GaN high electron mobility transistors (HEMT) coated by Lead ion selective membrane (Pb-ISM) was used (Chen et al. 2018). While using a HEMT, when the spacing in the gate electrode to HEMT layer was minimum, a large electric field was produced which further modulated the creation of EDL over the layer, and therefore, with respect to the ion concentration, the metal binding to the ISM was modified and thus, changed the transistor gain current. This device showed a high-sensitivity of −36 mV/log $[Pb^{2+}]$ with a LOD of 10^{-10} M. This study did not report the long-term stability of the device.

Another lead ion sensor based on a thermally treated rGO channel coated with the Al_2O_3 and Au NPs/ glutathione (GSH) for Pb^{2+} ions was reported (shown in Figure 6.2) (Maity et al. 2017). Sensor showed a fast response (~1–2 s) and LOD of a limit of < 1 ppb and high selectivity to lead ions.

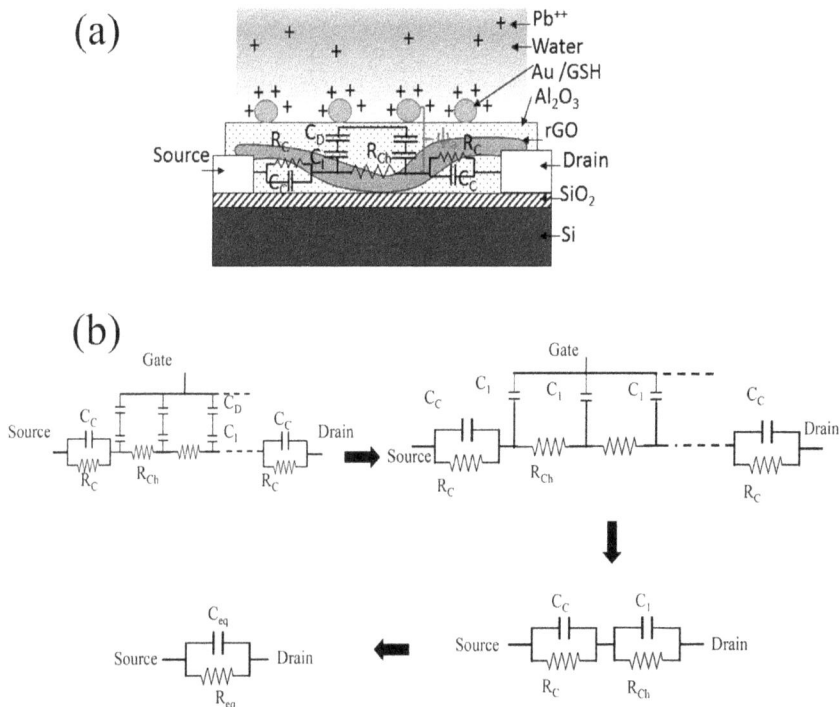

Figure 6.2 (a) Simplified model of insulated GFET based on Au/GSH for Pb^{2+} ion sensing, (b) Equivalent circuit model of the FET structure. [Reproduced with permission (Maity et al. 2017). Copyright 2017, American Chemical Society].

Apart from this, the graphene material was also widely explored for the detection of ions due to its sensitive surface as it is sensitive to variations in surface charge-carriers, or interface to ionic adsorbables (Li et al. 2017). However, for ion binding, the selective membrane must be used. In this field, a sensitive layer of Valinomycin is a neutral cyclic dodecadepsipeptide antibiotic which was used as a selective membrane and successfully measured the potassium ion concentration. The sensor was highly sensitive over 1 μm to 20 mm and obtained a response of 61 ± 4.6 mV/decade with a great stability over two months. However, the selectivity of the device was only tested to calcium and sodium ions. The fast, reasonable parallel detection and ultrasensitive label-free monitoring of potassium ions using AlGaN/GaN HEMTs were performed (Liu et al. 2020). The GaN acted as a stable material and was sensitive to surface charge movements and worked without a reference electrode. This device AlGaN/GaN HEMTs showed good sensitivity to K^+ ion concentration of 100 mm to 1 μm, with the response of 4.94 μA/lgαK^+, and stability of 28 days. Carbon nanotubes were also explored for ion sensing applications. The CNTs are very attractive in wear-based circuitry because of their good elasticity, charge-carrier movement, chemically stable, and ease of functionality. Recently, the CNTs ISFETs sodium ion detection was performed. In this work, the Ag/AgCl with polyvinyl butyral film acted as a reference electrode and the CNTs surface bonded PVC with sodium choosy film was used as a sensing electrode to capture Na^+ ions over 0.1 to 100 mM (Park et al. 2021). The sensor exhibited a high response of ~71.7 mV/dec to sodium chloride liquids. However, the reference electrode should maintain a steady voltage in different ion ranges. The di-(2-picolyl)amine (Dpa) derived based FETs were fabricated for copper ion sensing (Kenaan et al. 2020). The lipid single-layer acted as a top-gate dielectric material. DPA acted as a chelator which presented a great selective nature to Cu^{2+}. The lipid cover stopped trapped ions and thus, enhanced stability. The sensor exhibited good performance with a Nernstian slope of around 100 mV/ decade with LOD of 10 femtomolar. But the stability study of only 4 days was reported. Following Table 6.4 shows the development in the field of FETs based gas sensors

Table 6.4 Development of a few field effect transistors-based ion sensors

Materials used	Ion analyte	Sensor sensitivity	Limit of detection (LOD)	References
MoS_2/Au	Hg^{2+}	0.64 µA/ppb	11.52 ppt	(Nigam et al. 2020)
MPA-GSH	Cd^{2+}	0.241 µA/ppb	255 ppt	(Nigam et al. 2019a)
MoS_2	Cd^{2+}	~0.6	5 ng/mL	(Li et al. 2019)
DMTD	Pb^{2+}	0.607 µA/ppb	18 ppt	(Nigam et al. 2019b)
AlGaN/GaN	Pb^{2+}	-36 mV/log $[Pb^{2+}]$	10^{-10} m	(Chen et al. 2018)
Graphene ISFET	K^+	-63 mV/dec	–	(Li et al. 2017)
SiO_2/Si_3N_4-based transistor	K^+	-48 mV/dec		
AlGaN/GaN-based transistor	K^+	4.94 µA/lgαK$^+$	–	(Liu et al. 2020)
PEDOT/MWCNT/cotton fibers	K^+	–	1 nM	(Wang et al. 2022c)
CNT	Na^+	71.7 mV/dec	–	(Park et al. 2021)
OECT with PSSNa electrolyte	Na^+	~85 mV dec–1	10×10^{-6} m	(Han et al. 2020)

(Contd.)

Table 6.4   Development of a few field effect transistors-based ion sensors (*Contd.*)

Materials used	Ion analyte	Sensor sensitivity	Limit of detection (LOD)	References
SiO_2/Si_3N_4-based transistor	Na^+	–57 mV/dec	–	(Zhang et al. 2019)
SiO_2/Si_3N_4-based transistor	Ca^{2+}	–26 mV/dec		
ZnO@rGO	Cu^{2+}	185.32 mAμM^{-1}. cm^{-2}	14.9 μm	(Kim et al. 2022)
Dual-gated ISFET	Cu^{2+}	98 mV dec–1	10 fM	(Kenaan et al. 2020)
ZnO@GO	Cr^{3+}	49.28 mAμm^{-1}cm^{-2}	7.05 μm	(Kim et al. 2022)
Fluoropolysiloxane (FPSX) polymer-based matrix, nonactin and tetradodecylammonium nitrate (TDDAN)	NO^{3-}	~ 56 mV/decade	–	(Joly et al. 2022)

6.4   OPTICAL FIBER-BASED ION SENSORS

Optical fiber sensors have many merits such as small-size, anti-electromagnetic interference, chemically inert, and real-time monitoring ability, and therefore, an ideal sensing platform for the detection of heavy metal ion concentration. Generally, an optical fiber surface is coated with metal ion sensitive material. When the refractive index of the sensing material changes with respect to a change in ion concentration, the properties of light like wavelength, intensity, phase, or polarization state also changes as it is the function of heavy metal ion concentration. Further, the signal demodulation is done by a photodetector and electronic circuitry and the output signal is measured. A summary of some optical fiber ion sensors is summarized in Table 6.5.

The D-shaped polymeric type optical fiber based on an evanescence principle for mercury ion sensing was performed (Zhong et al. 2018). The self-assembled films of polycation (tris[2-(4-phenyldiazenyl) phenylaminoethoxy] cyclotriveratrylene (TPC) with poly dimethyl diallyl ammonium chloride (PDDAC)) and polyanion (TPC plus polyacrylic acid (PAA)) was employed for the detection purpose. This sensor showed a response time of 50 s to 0.05–2.5 mg/L mercury concentration with a LOD ~ 0.1 mg/L. Device response was not influenced by the addition of other ions like Cu^{2+}, Fe^{3+}, and Al^{3+}. Even though, sensor long-term stability needs to be considered for practical applications. The taper D fiber structure for proper handling is also necessary. Another mercury

Table 6.5 A summary of a few optical fiber ion sensors

Optical fiber type	Materials used	Ion analyte	Detection range	Sensor sensitivity	Limit of detection (LOD)	References
D-type fiber	PAA/TPC	Hg^{2+}	0.05–2.5 mg L^{-1}	–	0.1 mg/L	(Zhong et al. 2018)
Reflective SMF-NCF-FBG	Chitosan/PAA	Hg^{2+}	0–100 μm	0.0823 nm/μm	–	(Zhang et al. 2018)
SPR Uncladded fiber	Mercaptopyridine-functionalized gold nanoparticles	Hg^{2+}	8–100 nM	–	3.34 nm	(Yuan et al. 2019a)
TFG based fiber	Black phosperous	Pb^{2+}	0.1 ppb – 1.5×10^7 ppb	0.104 dB/μm	1.2 nm	(Liu et al. 2018)
Uncladded FBG	MA-AuNP	Pb^{2+}	10 fM – 100 nM	–	10 fM	(Kumar N et al. 2022)
Quartz optical fibers	Rhodizonate	Pb^{2+}	10 – 100 μg L^{-1}	0.1412 nm ppb^{-1}	85 ng/L	(Meza López et al. 2021)
Tapered SMF	Black phosphorus	Pb^{2+}	0.1 – 10^5 ppb	0.03714 nm/ppb	20.6 ppt	(Teng et al. 2021)
MZI based	Hydrogel	Pb^{2+}	2×10^{-7} mol/L – 1.2×10^{-6} mol/L	8.155×10^5 nm/ (mol/L)	2.452×10^{-8} mol/L	(Li et al. 2022)
SPR based uncladded multimode POF	Au NPs	Pb^{2+}	1 – 20 ppb	0.415 mV/ppb	1.2 ppb	(Boruah and Biswas 2018)
Etched FBG	ATAC	Cr^{2+}	–	722.8 pm/μm	–	(Kishore et al. 2018)

(Contd.)

Table 6.5 A summary of a few optical fiber ion sensors (*Contd.*)

Optical fiber type	Materials used	Ion analyte	Detection range	Sensor sensitivity	Limit of detection (LOD)	References
Uncladded MMF	CdTe QDs	Fe^{3+}	0 – 3.5 μm	–	14 nm	(Zhou et al. 2018)
No core fiber (NCF)	Chitosan/PAA	Ni^{2+}	0 – 500 μm	0.05537 nm/μm	0.1671 μm	(Raghunandhan et al. 2016)
SPR based unclad multimode PCS	Ag/CNT/Cu	NO_3^-	10^{-6} m – 5×10^{-3} m	80.62×10^6 nm/M	4 nM	(Parveen et al. 2017)
Silica microfiber	MWCNTs	Mg^{2+}	0.1 – 0.5%	23.27 dBm/%	0.0239%	(Yasin et al. 2018)
SPR based	ZnO/PANI coated on unclad MMF	HCO_3^-	0 – 200 μg/l	0.065 nm/(μg/l)	–	(Tabassum and Gupta 2016)
Microfiber with hairpin shape	Nucleic acid	Ag^+	–	0.22 nm/lg(m)	1.36 nM	(Yu et al. 2017)
SPR Uncladded MMF	Dicyclohexeno-18-crown-6 ionophore incorporated PVC -pH sensitive dye chromoionophore-I	K^+	80 – 2400 μm	–	80×10^{-6} m	(Sharma and Gupta 2021)
SPR based	CoO	$H_2PO_4^-$	0-100 ppm	0.3155 nm/ppm	4.96 ppm	(Verma and Gupta 2021)
Plastic optical fiber (POF)	Lumogallion-doped hydrogel	Al^{3+}	0-35 mm	0.594 dB/cm	–	(Chu et al. 2021)
Micro-tapered fiber LPGs	PDA-GO	Co^{2+}	$1-10^7$ ppb	–	<1 ppb	(Kang et al. 2022)

sensor based on FBG integrated with materials coated no-core fiber (NCF) (shown in Figure 6.3b) was fabricated (Zhang et al. 2018). By using the LBL technique, the layer of chitosan (CS)/poly acrylic acid (PAA) was layered over the NCF area. For good reflection of light, a silver film was also coated. When the sensing region was exposed to mercury concentration, the sensing material interacted with the mercury and thus, led to modification in the refractive index of the atrial and so, changes in the interference dip of the spectrum was observed. Additionally, the Bragg wavelength of FBGs was used to measure the surrounding temperature. The device showed a response around 0.0823 nm/μm to mercury over 0 μm–100 μm level and ~0.0178 nm/μm to 100 μm–500 μm, correspondingly. It showed a temperature response ~0.0147 nm/°C during 20 °C–50 °C. But, controlling coating layer thickness is very much important as it may affect the sensitivity of the sensor.

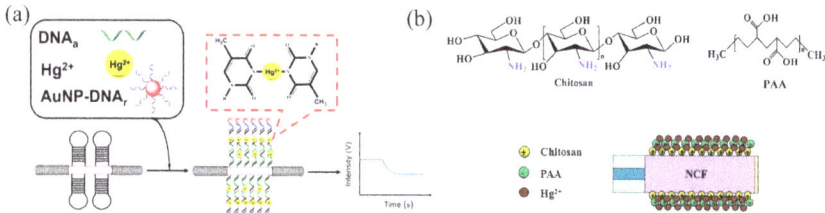

Figure 6.3 (a) Optical fiber mercury ion sensor based on immobilized hairpin probe (DNAmb), free supported probe (DNAa), the T–Hg^{2+}–T coordination chemistry, and a signal reporter based on an AuNP-labeled reporting probe (AuNP-DNAr). [Reproduced with permission (Fan et al. 2021). Copyright 2021, American Chemical Society]. (b) Optical fiber-based mercury ion sensor based on CS and PAA. [Reproduced with permission (Zhang et al. 2018). Copyright 2018, Elsevier].

At present, the tilted fiber grating (TFG) is fascinating because of its ability to couple light via core to the clad through an evanescence (Liu et al. 2018). The evanescence waves are very sensitive due to their weak guidance, and therefore, when they interact with the ion concentration, the wave gets affected by the change in ion concentration. Certain materials like black phosphorus (BP) have also proved to be very effective for ion sensing because of the high area for adsorption, large hole movement, and large adsorption energy. The sensitive material layered over the grating part of optical-fiber and the corresponding lead concentration range of 00.1 ppb to 1.5×10^7 ppb was monitored. The sensor performance was measured in power loss and obtained a sensitivity of 0.5×10^{-3} dB/ppb with a LOD of 0.25 ppb. The sensor stability and time characteristics need to be mentioned. The localized surface plasmon-based ion sensors are also very effective for ion sensing. They offer high sensitivity and reliability as the shift in the resonance dip is because of modification in material refractive index. In this, the SPR-dependent lead ion sensor is fabricated by using a U shape optical fiber (Boruah and Biswas 2018). The stripped part of an optical fiber was coated using the oxalic acid functionalized AuNPs and dipped in varied lead ion concentration. The ppb level LOD sensor in the linear detection range of 1–20 ppb was reported. However, due to the bending of the fiber surface, the power loss must be considered. The fluorescence materials-based sensors are very much promising due to their high quantum efficiency, great stability, and large band-width (Zhou et al. 2018). However, aimed at rapid on-site detection, the possible way to make a sensor is by mixing sensitive layer in a suitable matrix. In detail, the fluorescence-based CdTe quantum dots (QDs) doped hydrogel sensor for the detection of Fe^{3+} ions were reported. The QDs layered with N-Acetyl-l-cysteine (NAC) showed red emission with good selectivity and fluorescent reduction to Fe^{3+}. The sensor exhibited good linearity to 0–3.5 μm with a LOD of 14 nM. The MWCNTSs are also explored for ion detection due to their good electrical properties and sensitive nature to charge transfer effects on their surface by various analytes (Yasin et al. 2018). In this domain, the MWCNTs were drop-casted onto the silica microfiber and used to detect the magnesium ion concentration. The silica microfiber was used because it offered evanescent fields, constricted optical confinement, and manageable waveguide dispersion. The sensing mechanism was dependent upon the change in the refractive index of MWCNTs when it was exposed to magnesium solution. This work did not report the cross-selectivity study.

6.5   FLUORESCENCE BASED ION SENSORS

Fluorescent probes based on aggregation-induced emission (AIE) properties have received significant attention due to their good sensitivity, short response and recovery times low detection limits, high fluorescence quantum yields in their nano aggregated states, easy fabrication, use of moderate conditions, and selective recognition of organic/inorganic compounds in water with obvious changes in fluorescence. Table 6.6 shows a summary of fluorescence-based ion sensors.

The poly(NIPAM-co-TPE-SP), comprising N-isopropylacrylamide (NIPAM) along with tetraphenylethylene–spiropyran monomer (TPE-SP) for CN^- ion sensing was studied (Nhien et al. 2020). The fluorescent green TPE and red MC emissions at wavelengths of 517 and 627 nm, through Förster resonance energy transfer (FRET) was observed. One advantage of this material was that due to its biocompatibility, it can be used for cellular imagery and CN⁻ sensing in live cells. The light radiation at 627 nm increased slowly after adding water and thus, showed a characteristic AIE outcome. Another cyanide ion detection based on AIE used polysulfates via a Sulfur (VI) fluoride-based reaction was reported (Wan et al. 2020). It was observed that the material could measure CN⁻ selectively which was dependent upon the effective collaboration of anionic π to cyanide and naphthylamide rings. Due to the feeble chemical relations among other anionic and sensory, the device was highly selective to cyanide ions only. DNAs are also used in ion sensing applications. For example, a fluorescent-based lead ion sensor using C-PS2.M-DNA used Ag NCs was fabricated (Zhang and Wei 2018). When ions like lead ion come in contact with DNA, it changes aptamer and marks two dark DNA to Ag NCs nearer thus, further increasing the light intensity of Ag NCs and so, selective detection of lead ions occurs. The sensor detected a lead ion over 5 to 50 nM and LOD of 3.0 nM. The lead ions detection in Milli-Q water, tap water, or pond water was also performed successfully. A three-dimensional (3D) rotary paper–based microfluidic chip platform was also explored for lead ion sensing. In this, the fluorescent ZnSe quantum dots (QDs) with ion imprinting technique was used to monitor lead ion with a concentration range from 1 to 60 µg/L (Zhou et al. 2020). Generally, as that silicon wafers and glass, paper-based chips are cheap, ease in fabrication, small size, easy to carry and transport, and environmental user-friendly. The LOD of around 0.335 µg/L was reported. However, the sensor showed some quenching effect to cadmium ions as well. The mercury ion detection in water using tetraphenyl ethylene derivatives was reported. The PL at 477 nm decreased with the surge in mercury ions (0–100 nM) (Zhao et al. 2020). The corresponding PL changed from bright sky bluish to dark blue as shown in Figure 6.4(a). The real lake water sample was also tested by this system and thus, showed its use in real-time monitoring applications. The SCN⁻ detection in water was performed by using a phenanthrol[9,10-d] imidazole material (Bu et al. 2019). When water was added to DMF, the materials signal intensity gradually increased suggesting it followed a AIE signature. Due to the effective bonding of SCN⁻ to Ag^+, the PL emission was recovered after the addition of SCN⁻ to the phenanthrol [9,10-d] imidazole - Ag^+ system. The LOD of 7.8 nM in water was achieved by this system. A pyrene material was utilized for fluorine ion detection. The solvent was used as a CH_3CN/water mixture (Yadav et al. 2019). The sensing mechanism consisted of a steady creation of hydrogen bonds among the molecule and F⁻ at a minimum level and occurring deprotonation at higher level in CH_3CN. This sensor exhibited a good linear response to 1.99 µm–13.8 µm F⁻ concentration with LOD of 0.202 µm was observed.

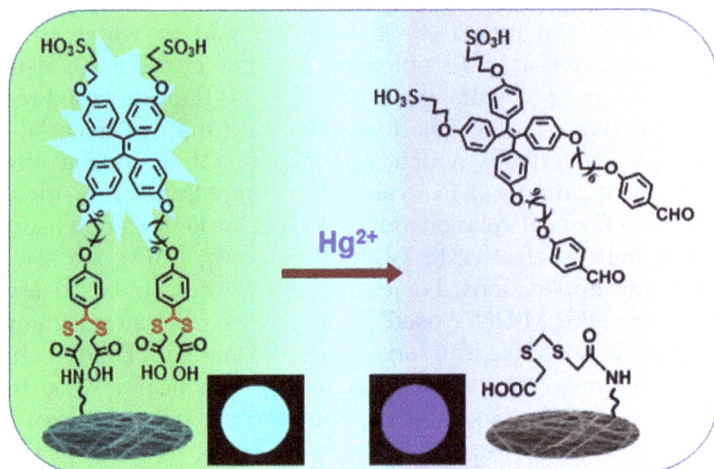

(a)

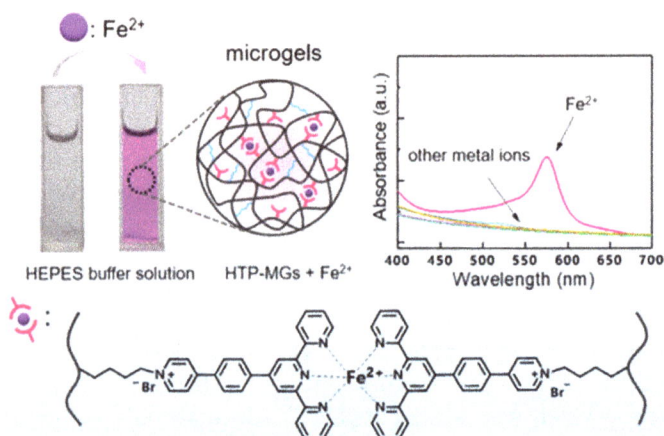

(b)

Figure 6.4 (a) Graphic of TPE-II grafted on electrospun fibers for Hg^{2+} sensing. [Reproduced with permission (Zhao et al. 2020). Copyright 2018, Elsevier]. (b) Possible binding structure of HTP moieties and Fe^{2+} ions and the fluorescence spectra of HTP-MG HEPES to Fe^{2+} ions. [Reproduced with permission (Ji et al. 2021). Copyright 2021, American Chemical Society].

Table 6.6 A summary of a few fluorescence-based ion sensors

Materials used	Ion analyte	Detection range (μM)	LOD (μM)	References
Poly(NIPAM-co-TPE-MC)	CN^-	0–80	–	(Nhien et al. 2020)
Polysulfates via a sulfur(vi) fluoride exchange (SuFEx)	CN^-	50–250	–	(Wan et al. 2020)
Cu@GSH	Pb^{2+}	200–700	106	(HAN et al. 2017)
Ag@C-PS2.M-DNA-templated	Pb^{2+}	5–50	–	(Zhang and Wei 2018)
ZnSe QDs	Pb^{2+}	1–60 μg/l	–	(Zhou et al. 2020)
Tetraphenylethene derivatives	Hg^{2+}	0–0.1	0.020	(Zhao et al. 2020)
phenanthro[9,10-d]imidazole	SCN^-	0–0.2	0.008	(Bu et al. 2019)
spirobifluorene-based probes	Zn^{2+}	0–100	0.019	(Wan et al. 2019)
Tetraphenylethylene Dimethylformamidine	PO_4^{3-}	–	–	(Yuan et al. 2022b)
thiourea-bridging bis-tetraphenylethylene	Cd^{2+}	0–0.16	0.236	(Jiang et al. 2020)
p-phenylenediamine with salicylaldehyde derivatives	AsO_4^{3-}	0–90	0.005	(Wang et al. 2019)
coumarin-salicylidene Schiff based probe 3	Cu^{2+}	0–30	0.024	(Padhan et al. 2019)
TPE-Sp-CN	HSO_3^-	0–500	10	(Lin et al. 2019)
Pyrene-thiophene based probe	F^-	1.99–13.5	435	(Yadav et al. 2019)

6.6 COLORIMETRIC ION SENSORS

The colorimetric method is very much used for qualitative and quantitative detection of ions and ionic mixtures. It is very simple, cost-effective, allows for naked eye observation, and there is no need of any sophisticated instrumentation. Certain features of this technique include on-site detection, fast, simple to fabricate, good reversibility, selectivity, and sensitivity, and therefore, adaptable in the research domain for speedy analysis. The colorimetric-based ion sensor is shown in Figure 6.4(b) and the summary of some ion sensors is tabulated in Table 6.7.

Another aluminum ion sensor was developed using rhodamine-based probe (R2PP) in $CH_3CN/DMSO$ (Kim et al. 2016). The sensing material was modified to R2PP mixed polyurethane electrospun (ES)

fibers shape. The rhodamine used was due to its good extinction coefficient, maximum quantum gain, and absorption constant. The intensity peak at 536 nm denoted a color variation from neutral to pink. The sensor showed a LOD of 8.5×10^{-9} m in the linear concentration of 0.05 to 2.5 μm. A colorimetric dependent Schiff base 2,3-dimethyl-4-(2-oxo-1,2-diphenylethylideneamino)-1-phenyl-1,2-dihydropyrazol-5-one (AP2) using chromium ion sensor detection was performed (Kim et al. 2016). The peak wavelength position shifted from 331 nm to 258 nm, which corresponded from yellow to colorless as observed easily by the naked eye. However, in the selectivity study, the sensor was also able to detect aluminum ions which might limit its use in a mixed sample. A carbazole moiety and 2-(((pyridin-2-ylmethyl)amino)methyl)phenol assembly utilizing an iron sensor was reported (Park et al. 2016). The sensing material had a lot of benefits for iron ion detection such as large intramolecular charge transmission and flexibility for structural adaptation. When it interacted with iron ions, its color changed from neutral to bright brownish to yellow. The absorption spectrum showed the sensor also responded to copper ions along with iron ions.

Table 6.7 A summary of a few colorimetric-based ion sensors

Materials used	Ion analyte	Solvent	LOD (μM)	References
Rhodamine-based probe (R2PP)	Al^{3+}	$CH_3CN:DMSO$ (2:8)	0.0085	(Kim et al. 2016)
2-Hydroxy-1-naphthaldehyde-(2-pyridyl) hydrazone (HL)	Al^{3+}	$H_2O:CH_3OH$ (9:1, pH = 5.3)	0.036	(Yu et al. 2016)
Polymerizable rhodamine monomer (GRBE)	Cr^{3+}	Neutral aqueous solution	2.20×10^{-6}	(Geng et al. 2016)
2,3-dimethyl-4-(2-oxo-1,2-diphenylethylideneamino)-1-phenyl-1,2-dihydropyrazol-5-one (AP2)	Cr^{3+}	–	–	(Kim et al. 2016)
Carbazole moiety and 2-(((pyridin-2-ylmethyl) amino)methyl)phenol group	Fe^{3+}	CH_3OH:bis-tris buffer (1:1)	13.5	(Park et al. 2016)
imine boronic esters functionalized with pyridine	Co^{2+} Cu^{2+}	– –	– –	(Sánchez-Portillo et al. 2021)
p (AAc/AMPS)-TA hydrogel	Cr^6 Mn^{2+} Ni^{2+} Co^{2+}	– – – –	5×10^{-4} m	(El-damhougy et al. 2021)

6.7   CONCLUSION AND OUTLOOK

In this chapter, progress in recent years on ion sensing based on various techniques is reported. Numerous methods including electrochemical, field effect transistors, fiber optics, fluorescence, and colorimetry are summarized and discussed. Their importance for ion detection, principle, material role, merits, and demerits are provided. The advanced fabrication coating methods, efficient materials synthesis methods, and fine integration of these materials with these techniques have greatly enhanced the sensing performance for the detection of these toxic metal ions. The discussion regarding the metal ion sensing performance of these techniques, key parameters, and future scope is summarized below.

1.   Electrochemical sensors are extensively used for metal ion sensing due to their inexpensive and time-efficient. The potentiometric based ion sensing is considerably used due to its ability to achieve sufficiently lower detection limits, cheap, and ease in preparation. However, the selectivity needs to be studied. Tuning ISM composition is needed to obtain the high selectivity and also it could address the other interfering parameters. Certain parameters such as water accumulation at the membrane, the presence of gases in the solution, and changes in pH and temperature could affect the performance of the sensor, and therefore, need to be considered. The sensor should sustain longer times at the environmental conditions, and therefore, new materials for solid-contacts by using conducting polymers and carbon-based nanomaterials are highly preferred. Also, electrochemical methods require prior chemical treatment, and therefore, there might be a chance of contamination.

2.   FETs are thoroughly investigated for ion sensing. Lot many MOXs, 2D-materials, and polymers have been extensively used for ion sensing. Most of the devices showed a higher Nernstian slope and also a lower LOD. Yet, the sensor's stability study and effect on sensing mechanism need to be focused.

3.   Numerous optical fiber sensors have been investigated for ion detection. Different kinds of optical fiber geometries like tapered fiber, unclad fibers, U-bend, D-type, and microfiber geometries are integrated with various metal oxides, metals, DNA functionalized, carbon-based materials, polymers, and their composites have been studied for ion detection. The tapered fiber has shown good sensitivity but it decreases the mechanical strength. A lot of metal-coated SPR sensors were observed but it is a complex process and for this purpose, high uniformity in the

metallic coating is always required. The precise binding of ions to the sensitive layer is also important. The interfering effect by other ions is still required. Most of the sensors are reported the detection of one ion only. The film thickness is very important for ion sensing. So, the film thickness needs to be controlled for effective interaction between the ion and sensing layer. The reusable ion sensor must be developed. The effect of temperature and humidity on the sensing layer must be controlled.

4. The colorimetry method is also used for effective ion sensing. The advantages include fast detection, easy monitoring, low-cost, easy preparation, etc. The unique advantage of this method is the simultaneous detection of many metal ions in an array. However, the chemical reactions might be irreversible, and therefore, can be used for a single time only. Thus, this method is not much ideal for continuous monitoring applications.

5. A sensing material combined with multifunctional methods for ion sensing is very necessary.

REFERENCES

Adarakatti, Prashanth Shivappa, Ashoka Siddaramanna and Pandurangappa Malingappa. 2019. Fabrication of a new Calix[4]arene-functionalized Mn_3O_4 nanoparticle-based modified glassy carbon electrode as a fast responding sensor towards Pb^{2+} and Cd^{2+} ions. Analytical Methods 11(6): 813–820. The Royal Society of Chemistry. doi:10.1039/C8AY02648C.

Alizadeh, Taher, Faride Rafiei, Negin Hamidi and Mohamad Reza Ganjali. 2017. A new electrochemical sensing platform for Cr(III) determination based on nano-structured Cr(III)-imprinted polymer-modified carbon composite electrode. Electrochimica Acta 247: 812–819. doi:https://doi.org/10.1016/j.electacta.2017.07.081.

Alizadeh, Taher and Khalil Atayi. 2018. Synthesis of nano-sized hydrogen phosphate-imprinted polymer in acetonitrile/water mixture and its use as a recognition element of hydrogen phosphate selective all-solid state potentiometric electrode. Journal of Molecular Recognition 31(2): e2678. John Wiley & Sons, Ltd. doi:https://doi.org/10.1002/jmr.2678.

Altuner, Elif Esra, Veli Cengiz Ozalp, M. Deniz Yilmaz, Mert Sudagidan, Aysenur Aygun, Elif Esma Acar, Behiye Busra Tasbasi and Fatih Sen. 2022. Development of electrochemical aptasensors detecting phosphate ions on TMB substrate with epoxy-based mesoporous silica nanoparticles. Chemosphere 297: 134077. doi:https://doi.org/10.1016/j.chemosphere.2022.134077.

Boruah, Bijoy Sankar and Rajib Biswas. 2018. Localized surface plasmon resonance based U-shaped optical fiber probe for the detection of Pb^{2+} in aqueous medium. Sensors and Actuators B: Chemical 276: 89–94. doi:https://doi.org/10.1016/j.snb.2018.08.086.

Bu, Fanqiang, Bing Zhao, Wei Kan, Liyan Wang, Bo Song, Jianxin Wang, Zhe Zhang, et al. 2019. A phenanthro[9,10-d]imidazole-based AIE active fluorescence probe for sequential detection of Ag^+/AgNPs and SCN^- in water and saliva samples and its application in living cells. Spectrochimica Acta Part A: Molecular and Biomolecular Spectroscopy 223: 117333. doi:https://doi.org/10.1016/j.saa.2019.117333.

Chen, Yi-Ting, Indu Sarangadharan, Revathi Sukesan, Ching-Yen Hseih, Geng-Yen Lee, Jen-Inn Chyi and Yu-Lin Wang. 2018. High-field modulated ion-selective field-effect-transistor (FET) sensors with sensitivity higher than the ideal nernst sensitivity. Scientific Reports 8(1): 8300. doi:10.1038/s41598-018-26792-9.

Chu, Fenghong, Pengfei Han, Shi Feng, Shuangjiao Wei, Huyong Ma and Zhenglan Bian. 2021. Hydrogel optical fibers functionalized with lumogallion as aluminum ions sensing platform. Optik 240: 166875. doi:https://doi.org/10.1016/j.ijleo.2021.166875.

Ebrahimi, Nastaran, Jahan Bakhsh Raoof, Reza Ojani and Maryam Ebrahimi. 2022. Designing a novel DNA-based electrochemical biosensor to determine of Ba^{2+} ions both selectively and sensitively. Analytical Biochemistry 642: 114563. doi:https://doi.org/10.1016/j.ab.2022.114563.

Echabaane, M., A. Rouis, M.A. Mahjoub, I. Bonnamour and H. Ben Ouada. 2017. Impedimetric sensing proprieties of ITO electrodes functionalized with PEDOT:PSS/Azo-Calix[4] arene for the detection of Al^{3+} ions under light excitation. Journal of Electronic Materials 46(1): 418–424. doi:10.1007/s11664-016-4838-1.

El-damhougy, Tasneam K., Amal S.I. Ahmed, Ghalia A. Gaber, Nabila A. Mazied and Ghada Bassioni. 2021. Radiation synthesis for a highly sensitive colorimetric hydrogel sensor-based p(AAc/AMPS)-TA for metal ion detection. Results in Materials 9: 100169. doi:https://doi.org/10.1016/j.rinma.2021.100169.

Fan, Shu-Mei, Chang-Yue Chiang, Yen-Ta Tseng, Tsung-Yan Wu, Yen-Ling Chen, Chun-Jen Huang and Lai-Kwan Chau. 2021. Detection of Hg(II) at part-per-quadrillion levels by fiber optic plasmonic absorption using DNA hairpin and DNA-gold nanoparticle conjugates. ACS Applied Nano Materials 4(10): 10128–10135. American Chemical Society. doi:10.1021/acsanm.1c01566.

Geng, Tong-Mou, Chang Guo, Yan-Jie Dong, Meng Chen and Yu Wang. 2016. Turn-on fluorogenic and chromogenic detection of cations in complete water media with Poly(N-Vinyl Pyrrolidone) bearing rhodamine B derivatives as polymeric chemosensor. Polymers for Advanced Technologies 27(1): 14390–14397. John Wiley & Sons, Ltd. doi:https://doi.org/10.1002/pat.3603.

Ghanei-Motlagh, M. and M.A. Taher. 2017a. Magnetic silver(I) ion-imprinted polymeric nanoparticles on a carbon paste electrode for voltammetric determination of silver(I). Microchimica Acta 184(6): 1691–1699. doi:10.1007/s00604-017-2157-8.

Ghanei-Motlagh, M. and M.A. Taher. 2017b. Novel imprinted polymeric nanoparticles prepared by Sol–Gel technique for electrochemical detection of toxic Cadmium(II) ions. Chemical Engineering Journal 327: 135–141. doi:https://doi.org/10.1016/j.cej.2017.06.091.

Guin, Saurav K., Arvind S. Ambolikar, Jhimli Paul Guin and Suman Neogy. 2018. Exploring the excellent photophysical and electrochemical properties of graphene quantum dots for complementary sensing of uranium. Sensors and Actuators B: Chemical 272: 559–573. doi:https://doi.org/10.1016/j.snb.2018.05.176.

Han, Bing-Yan, Xu-Fen Hou, Rong-Chao Xiang, Ming-Bo Yu, Ying Li, Ting-Ting Peng and Gao-Hong He. 2017. Detection of lead ion based on aggregation-induced emission of copper nanoclusters. Chinese Journal of Analytical Chemistry 45(1): 23–27. doi:https://doi.org/10.1016/S1872-2040(16)60985-4.

Han, Sanggil, Shunsuke Yamamoto, Anastasios G. Polyravas and George G. Malliaras. 2020. Microfabricated ion-selective transistors with fast and super-nernstian response. Advanced Materials 32(48): 2004790. John Wiley & Sons, Ltd. doi:https://doi.org/10.1002/adma.202004790.

He, Qing, Suprem R. Das, Nathaniel T. Garland, Dapeng Jing, John A. Hondred, Allison A. Cargill, Shaowei Ding, et al. 2017. Enabling inkjet printed graphene for ion selective electrodes with postprint thermal annealing. ACS Applied Materials & Interfaces 9(14): 12719–12727. American Chemical Society. doi:10.1021/acsami.7b00092.

Hojatpanah, Mohammad Reza, Akbar Khanmohammadi, Hosin Khoshsafar, Ali Hajian and Hasan Bagheri. 2022. Construction and application of a novel electrochemical sensor for trace determination of uranium based on ion-imprinted polymers modified glassy carbon electrode. Chemosphere 292: 133435. doi:https://doi.org/10.1016/j.chemosphere.2021.133435.

Jaishankar, Monisha, Tenzin Tseten, Naresh Anbalagan, Blessy B. Mathew and Krishnamurthy N. Beeregowda. 2014. Toxicity, mechanism and health effects of some heavy metals. Interdisciplinary Toxicology 7(2). Slovak Toxicology Society SETOX: 60–72. doi:10.2478/intox-2014-0009.

Ji, Weiming, Shunni Dong, Wei Fan, Chao Lv, Jingjing Nie and Binyang Du. 2021. Functional microgel for selective and sensitive colorimetric detection of Fe^{2+} ions in HEPES buffer aqueous solutions. ACS Applied Polymer Materials 3(5): 2489–2497. American Chemical Society. doi:10.1021/acsapm.1c00073.

Jiang, Shengjie, Shibing Chen, Zhengchao Wang, Hongyu Guo and Fafu Yang. 2020. First fluorescence sensor for simultaneously detecting three kinds of IIB elements (Zn^{2+}, Cd^{2+} and Hg^{2+}) based on aggregation-induced emission. Sensors and Actuators B: Chemical 308: 127734. doi:https://doi.org/10.1016/j.snb.2020.127734.

Joly, M., M. Marlet, C. Durieu, C. Bene, J. Launay and P. Temple-Boyer. 2022. Study of chemical field effect transistors for the detection of ammonium and nitrate ions in liquid and soil phases. Sensors and Actuators B: Chemical 351: 130949. doi:https://doi.org/10.1016/j.snb.2021.130949.

Kang, Xin, Ruiduo Wang, Man Jiang, Erkang Li, Yarong Li, Xiaoxin Yan, Tianqi Wang, et al. 2022. Polydopamine functionalized graphene oxide for high sensitivity micro-tapered long period fiber grating sensor and its application in detection Co^{2+} ions. Optical Fiber Technology 68: 102807. doi:https://doi.org/10.1016/j.yofte.2021.102807.

Kenaan, A., F. Brunel, J.-M. Raimundo and A.M. Charrier. 2020. Femtomolar detection of Cu^{2+} ions in solution using super-nernstian FET-sensor with a lipid monolayer as top-gate dielectric. Sensors and Actuators B: Chemical 316: 128147. doi:https://doi.org/10.1016/j.snb.2020.128147.

Kim, Changkyeom, Ji-Yong Hwang, Kyo-Sun Ku, Satheshkumar Angupillai and Young-A Son. 2016. A renovation of non-aqueous Al^{3+} sensor to aqueous media sensor by simple recyclable immobilize electrospun nano-fibers and its uses for live sample analysis. Sensors and Actuators B: Chemical 228: 259–269. doi:https://doi.org/10.1016/j.snb.2016.01.020.

Kim, Eun-Bi, M. Imran, Eun-Hee Lee, M. Shaheer Akhtar and Sadia Ameen. 2022. Multiple ions detection by field-effect transistor sensors based on ZnO@GO and ZnO@rGO nanomaterials: application to trace detection of Cr (III) and Cu (II). Chemosphere 286: 131695. doi:https://doi.org/10.1016/j.chemosphere.2021.131695.

Kishore, Pabbisetti Vayu Nandana, Sai Shankar Madhuvarasu and Satyanarayana Moru. 2018. Stimulus responsive hydrogel-coated etched fiber bragg grating for carcinogenic chromium (VI) sensing. Optical Engineering 57(1): 1–7. doi:10.1117/1.OE.57.1.017101.

Kumar N., Vajresh, Kavitha B.S. and S. Asokan. 2022. Selective detection of lead in water using etched fiber bragg grating sensor. Sensors and Actuators B: Chemical 354: 131208. doi:https://doi.org/10.1016/j.snb.2021.131208.

Li, Hongmei, Yihao Zhu, Md. Sayful Islam, Md Anisur Rahman, Kenneth B. Walsh and Goutam Koley. 2017. Graphene field effect transistors for highly sensitive and selective detection of K^+ Ions. Sensors and Actuators B: Chemical 253: 759–765. doi:https://doi.org/10.1016/j.snb.2017.06.129.

Li, Peng, Dongzhi Zhang and Zhenling Wu. 2019. Flexible MoS_2 sensor arrays for high performance label-free ion sensing. Sensors and Actuators A: Physical 286: 51–58. doi:https://doi.org/10.1016/j.sna.2018.12.026.

Li, Gengsong, Zhen Liu, Jianxun Feng, Guiyao Zhou and Xuguang Huang. 2022. Pb^{2+} fiber optic sensor based on smart hydrogel coated mach-zehnder interferometer. Optics & Laser Technology 145: 107453. doi:https://doi.org/10.1016/j.optlastec.2021.107453.

Lin, Tingting, Xing Su, Kai Wang, Minjie Li, Hongwei Guo, Lulu Liu, Bo Zou, et al. 2019. An AIE fluorescent switch with multi-stimuli responsive properties and applications for quantitatively detecting PH value, sulfite anion and hydrostatic pressure. Materials Chemistry Frontiers 3(6): 1052–1061. The Royal Society of Chemistry. doi:10.1039/C8QM00544C.

Liu, C., Z. Sun, L. Zhang, J. Lv, X.F. Yu, L. Zhang and X. Chen. 2018. Black phosphorus integrated tilted fiber grating for ultrasensitive heavy metal sensing. Sensors and Actuators B: Chemical 257: 1093–1098. doi:https://doi.org/10.1016/j.snb.2017.11.022.

Liu, Yueling, Yang Liu, Yingying Gao and Ping Wang. 2019. A General approach to one-step fabrication of single-piece nanocomposite membrane based Pb^{2+}-selective electrodes. Sensors and Actuators B: Chemical 281: 705–712. doi:https://doi.org/10.1016/j.snb.2018.09.113.

Liu, Xinsheng, Lei Zhao, Bin Miao, Zhiqi Gu, Jin Wang, Huoxiang Peng, Jiande Li, et al. 2020. Wearable multiparameter platform based on AlGaN/ GaN high-electron-mobility transistors for real-time monitoring of PH and potassium ions in sweat. Electroanalysis 32(2): 422–428. John Wiley & Sons, Ltd. doi:https://doi.org/10.1002/elan.201900405.

Maity, Arnab, Xiaoyu Sui, Chad R. Tarman, Haihui Pu, Jingbo Chang, Guihua Zhou, Ren Ren, et al. 2017. Pulse-driven capacitive lead ion detection with reduced graphene oxide field-effect transistor integrated with an analyzing device for rapid water quality monitoring. ACS Sensors 2(11): 1653–1661. American Chemical Society. doi:10.1021/acssensors.7b00496.

Mariyappan, V., S. Manavalan, Shen-Ming Chen, G. Jaysiva, P. Veerakumar and M. Keerthi. 2020. Sr@FeNi-S nanoparticle/carbon nanotube nanocomposite with superior electrocatalytic activity for electrochemical detection of toxic mercury(II). ACS Applied Electronic Materials 2(7): 1943–1952. American Chemical Society. doi:10.1021/acsaelm.0c00248.

McCaul, Margaret, Adam Porter, Ruairi Barrett, Paddy White, Florien Stroiescu, Gordon Wallace and Dermot Diamond. 2018. Wearable platform for real-time monitoring of sodium in sweat. ChemPhysChem 19(12): 1531–1536. John Wiley & Sons, Ltd. doi:https://doi.org/10.1002/cphc.201701312.

Mei, Chong J., Nor A. Yusof and Shahrul A. Alang Ahmad. 2021. Electrochemical determination of lead & amp; copper ions using thiolated Calix[4]arene-modified screen-printed carbon electrode. Chemosensors. doi:10.3390/ chemosensors9070157.

Mendecki, Lukasz and Katherine A. Mirica. 2018. Conductive metal–organic frameworks as ion-to-electron transducers in potentiometric sensors. ACS Applied Materials & Interfaces 10(22): 19248–19257. American Chemical Society. doi:10.1021/acsami.8b03956.

Meza López, Flor de Liss, S. Khan, G. Picasso and Maria Del Pilar T. Sotomayor. 2021. A novel highly sensitive imprinted polymer-based optical sensor for the detection of Pb(II) in water samples. Environmental Nanotechnology, Monitoring & Management 16: 100497. doi:https://doi.org/10.1016/j.enmm. 2021.100497.

Mishra, Sandhya, Ram Naresh Bharagava, Nandkishor More, Ashutosh Yadav, Surabhi Zainith, Sujata Mani and Pankaj Chowdhary. 2019. Heavy metal contamination: an alarming threat to environment and human health. pp 103–125. In: R. Sobti, N. Arora and R. Kothari (eds). Environmental Biotechnology: For Sustainable Future. Springer, Singapore. doi:10.1007/978-981-10-7284-0_5.

Nhien, Pham Quoc, Wei-Lun Chou, Tu Thi Kim Cuc, Trang Manh Khang, Chia-Hua Wu, Natesan Thirumalaivasan, Bui Thi Buu Hue, et al. 2020. Multi-stimuli responsive FRET processes of bifluorophoric AIEgens in an amphiphilic copolymer and its application to cyanide detection in aqueous media. ACS Applied Materials & Interfaces 12(9): 10959–10972. American Chemical Society. doi:10.1021/acsami.9b21970.

Nigam, A., T.N. Bhat, V.S. Bhati, S.B. Dolmanan, S. Tripathy and M. Kumar. 2019a. MPA-GSH functionalized AlGaN/GaN high-electron mobility transistor-based sensor for cadmium ion detection. IEEE Sensors Journal 19(8): 2863–2870. doi:10.1109/JSEN.2019.2891511.

Nigam, A., V.S. Bhati, T.N. Bhat, S.B. Dolmanan, S. Tripathy and M. Kumar. 2019b. Sensitive and selective detection of Pb^{2+} ions using 2,5-Dimercapto-1,3,4-Thiadiazole functionalized AlGaN/GaN high electron mobility transistor. IEEE Electron Device Letters 40(12): 1976–1979. doi:10.1109/LED.2019.2947141.

Nigam, Adarsh, Neeraj Goel, Thirumaleshwara N. Bhat, Md. Tawabur Rahman, Surani Bin Dolmanan, Qiquan Qiao, Sudhiranjan Tripathy, et al. 2020. Real time detection of Hg^{2+} ions using MoS_2 functionalized AlGaN/GaN high electron mobility transistor for water quality monitoring. Sensors and Actuators B: Chemical 309: 127832. doi:https://doi.org/10.1016/j.snb.2020.127832.

Nigam, Adarsh, Nipun Sharma, Sudhiranjan Tripathy and Mahesh Kumar. 2021. Development of semiconductor based heavy metal ion sensors for water analysis: a review. Sensors and Actuators A: Physical 330: 112879. doi:https://doi.org/10.1016/j.sna.2021.112879.

Norouzi, Banafsheh and Zahra Parsa. 2018. Determination of sulfite in real sample by an electrochemical sensor based on Ni/Poly(4-Aminobenzoic Acid)/Sodium Dodecylsulfate/Carbon paste electrode. Russian Journal of Electrochemistry 54(8): 613–622. doi:10.1134/S1023193518080049.

Padhan, Subrata Kumar, Narayan Murmu, Subrat Mahapatra, M.K. Dalai and Satya Narayan Sahu. 2019. Ultrasensitive detection of aqueous Cu^{2+} ions by a coumarin-salicylidene based Aiegen. Materials Chemistry Frontiers 3(11): 2437–2347. The Royal Society of Chemistry. doi:10.1039/C9QM00394K.

Park, Gyeong Jin, Ga Rim You, Ye Won Choi and Cheal Kim. 2016. A naked-eye chemosensor for simultaneous detection of iron and copper ions and its copper complex for colorimetric/fluorescent sensing of cyanide. Sensors and Actuators B: Chemical 229: 257–271. doi:https://doi.org/10.1016/j.snb.2016.01.133.

Park, Sang-Chan, Hee June Jeong, Min Heo, Jae Ho Shin and Jae-Hyuk Ahn. 2021. Carbon nanotube-based ion-sensitive field-effect transistors with an on-chip reference electrode toward wearable sodium sensing. ACS Applied Electronic Materials 3(6): 2580–2588. American Chemical Society. doi:10.1021/acsaelm.1c00152.

Parveen, Shama, Anisha Pathak, and B.D. Gupta. 2017. Fiber optic SPR nanosensor based on synergistic effects of CNT/Cu-Nanoparticles composite for ultratrace sensing of nitrate. Sensors and Actuators B: Chemical 246: 910–919. doi:https://doi.org/10.1016/j.snb.2017.02.170.

Patella, B., G. Aiello, G. Drago, C. Torino, A. Vilasi, A. O'Riordan and R. Inguanta. 2022. Electrochemical detection of chloride ions using ag-based electrodes obtained from compact disc. Analytica Chimica Acta 1190: 339215. doi:https://doi.org/10.1016/j.aca.2021.339215.

Pizarro, Jaime, Erick Flores, Victor Jimenez, Tamara Maldonado, Claudio Saitz, Andres Vega, Fernando Godoy, et al. 2019. Synthesis and characterization of the first cyrhetrenyl-appended Calix[4] arene macrocycle and its application as an electrochemical sensor for the determination of Cu(II) in bivalve mollusks using square wave anodic stripping voltammetry. Sensors and Actuators B: Chemical 281: 115–122. doi:https://doi.org/10.1016/j.snb.2018.09.099.

Raghunandhan, R., L.H. Chen, H.Y. Long, L.L. Leam, P.L. So, X. Ning and C.C. Chan. 2016. Chitosan/PAA based fiber-optic interferometric sensor for heavy metal ions detection. Sensors and Actuators B: Chemical 233: 31–38. doi:https://doi.org/10.1016/j.snb.2016.04.020.

Rashed, Md. A., M. Faisal, F.A. Harraz, Md. Jalalah, Mabkhoot Alsaiari and M.S. Al-Assiri. 2020. RGO/ZnO/Nafion nanocomposite as highly sensitive and selective amperometric sensor for detecting nitrite ions (NO_2^-). Journal of the Taiwan Institute of Chemical Engineers 112: 345–356. doi:https://doi.org/10.1016/j.jtice.2020.05.015.

Ru, Jing, Xuemei Wang, Zheng Zhou, Jiali Zhao, Jing Yang, Xinzhen Du and Xiaoquan Lu. 2022. Fabrication of octahedral GO/UiO-67@PtNPs nanocomposites as an electrochemical sensor for ultrasensitive recognition of arsenic (III) in Chinese herbal medicine. Analytica Chimica Acta 1195: 339451. doi:https://doi.org/10.1016/j.aca.2022.339451.

Sánchez-Portillo, Paola, Aime Hernández-Sirio, Carolina Godoy-Alcántar, Pascal G. Lacroix, Vivechana Agarwal, Rosa Santillán and Victor Barba. 2021. Colorimetric metal ion (II) sensors based on imine boronic esters functionalized with pyridine. Dyes and Pigments 186: 108991. doi:https://doi.org/10.1016/j.dyepig.2020.108991.

Sharma, Sonika and Banshi D. Gupta. 2021. Surface plasmon resonance based fiber optic potassium ion disposable sensing probe for soil testing. Optical Fiber Technology 64: 102573. doi:https://doi.org/10.1016/j.yofte.2021.102573.

Shi, Haobing, Li Fu, Fei Chen, Shichao Zhao and Guosong Lai. 2022. Preparation of highly sensitive electrochemical sensor for detection of nitrite in drinking water samples. Environmental Research 209: 112747. doi:https://doi.org/10.1016/j.envres.2022.112747.

Sun, Xiaotong, Tanji Yin, Ziping Zhang and Wei Qin. 2022. Redox probe-based amperometric sensing for solid-contact ion-selective electrodes. Talanta 239: 123114. doi:https://doi.org/10.1016/j.talanta.2021.123114.

Tabassum, Rana and Banshi D. Gupta. 2016. Tailoring the field distribution of ZnO by polyaniline for SPR-based fiber optic detection of hardness of the drinking water. Plasmonics 11(2): 483–492. doi:10.1007/s11468-015-0079-z.

Teng, Pingping, Yuhan Jiang, Xinyu Chang, Yu Shen, Zhihai Liu, Nigel Copner, Jun Yang, et al. 2021. Highly sensitive on-line detection of trace Pb^{2+} based on tapered fiber integrated with black phosphorus. Optical Fiber Technology 66: 102668. doi:https://doi.org/10.1016/j.yofte.2021.102668.

Verma, Assim and Banshi D. Gupta. 2021. Fiber optic surface plasmon resonance based disposable probe for the detection of phosphate ion in soil. Optik 243: 167484. doi:https://doi.org/10.1016/j.ijleo.2021.167484.

Wan, Jianyong, Wu Zhang, Hongda Guo, JingJing Liang, Danyu Huang and Haibo Xiao. 2019. Two spirobifluorene-based fluorescent probes with aggregation-induced emission properties: synthesis and application in the detection of Zn^{2+} and cell imaging. Journal of Materials Chemistry C 7(8): 2240–2249. The Royal Society of Chemistry. doi:10.1039/C8TC05526B.

Wan, Haibo, Shiyuan Zhou, Peiyang Gu, Feng Zhou, Da Lyu, Qinghua Xu, Anna Wang, et al. 2020. AIE-active polysulfates via a sulfur(vi) fluoride exchange (SuFEx) click reaction and investigation of their two-photon fluorescence and cyanide detection in water and in living cells11electronic supplementary information (ESI) available. see DOI: 10.1039. Polymer Chemistry 11(5): 1033–42. doi:https://doi.org/10.1039/c9py01448a.

Wang, Lina, Yong Li, Xike Tian, Chao Yang, Liqiang Lu, Zhaoxin Zhou, Yunjie Huang, et al. 2019. Construction of salicylaldehyde analogues as turn-on fluorescence probes and their electronic effect on sensitive and selective detection of As(v) in groundwater. Analytical Methods 11(7): 955–964. The Royal Society of Chemistry. doi:10.1039/C8AY02484G.

Wang, Fei-Fei, Chang Liu, Jin Yang, Hong-Liang Xu, Wen-Yuan Pei and Jian-Fang Ma. 2022a. A sulfur-containing capsule-based metal-organic electrochemical sensor for super-sensitive capture and detection of multiple heavy-metal ions. Chemical Engineering Journal 438: 135639. doi:https://doi.org/10.1016/j.cej.2022.135639.

Wang, Ri, Chen-Yu Xiong, Yong Xie, Ming-Jie Han, Yu-Hao Xu, Chao Bian and Shan-Hong Xia. 2022b. Electrochemical sensor based on MoS_2 nanosheets and DNA hybridization for trace mercury detection. Chinese Journal of Analytical Chemistry 50(3): 100066. doi:https://doi.org/10.1016/j.cjac.2022.100066.

Wang, Yao, Yuedan Wang, Rufeng Zhu, Yang Tao, Yuanli Chen, Qiongzhen Liu, Xue Liu, et al. 2022c. Woven fiber organic electrochemical transistors based on multiwalled carbon nanotube functionalized PEDOT nanowires for nondestructive detection of potassium ions. Materials Science and Engineering: B 278: 115657. doi:https://doi.org/10.1016/j.mseb.2022.115657.

Xu, Gang, Chen Cheng, Wei Yuan, Zhaoyang Liu, Lihang Zhu, Xintong Li, Yanli Lu, et al. 2019. Smartphone-based battery-free and flexible electrochemical patch for calcium and chloride ions detections in biofluids. Sensors and Actuators B: Chemical 297: 126743. doi:https://doi.org/10.1016/j.snb.2019.126743.

Xu, Jinming, Huangmei Zhou, Yixue Zhang, Yu Zhao, Hao Yuan, Xiaoxiao He, Ying Wu, et al. 2022a. Copper nanoclusters-based fluorescent sensor array to identify metal ions and dissolved organic matter. Journal of Hazardous Materials 428: 128158. doi:https://doi.org/10.1016/j.jhazmat.2021.128158.

Xu, Kebin, Binyu Wu, Junliang Wan, Ying Li and Min Li. 2022b. A potentiometric phosphate ion sensor based on electrochemically modified nickel electrode. Electrochimica Acta 412: 140065. doi:https://doi.org/10.1016/j.electacta.2022.140065.

Yadav, Pranjalee, Sarita Gond, Ashish Kumar Singh and Vinod P. Singh. 2019. A pyrene-thiophene based probe for aggregation induced emission enhancement (AIEE) and naked-eye detection of fluoride ions. Journal of Luminescence 215: 116704. doi:https://doi.org/10.1016/j.jlumin.2019.116704.

Yasin, M., N. Irawati, N.M. Isa, S.W. Harun and F. Ahmad. 2018. MWCNTs coated silica microfiber sensor for detecting Mg^{2+} in de-ionized water. Optik 171: 65–70. doi:https://doi.org/10.1016/j.ijleo.2018.05.132.

Yin, Tanji, Tingting Han, Changbai Li, Wei Qin and Johan Bobacka. 2020. Real-time monitoring of the dissolution of silver nanoparticles by using a solid-contact Ag^+-selective electrode. Analytica Chimica Acta 1101: 50–57. doi:https://doi.org/10.1016/j.aca.2019.12.022.

150

Chemical Sensors

Yu, Fang, Ling Jie Hou, Li Yuan Qin, Jian Bin Chao, Yu Wang and Wei Jun Jin. 2016. A new colorimetric and turn-on fluorescent chemosensor for Al^{3+} in aqueous medium and its application in live-cell imaging. Journal of Photochemistry and Photobiology A: Chemistry 315: 8–13. doi:https://doi.org/10.1016/j.jphotochem.2015.09.006.

Yu, Bo, Yunyun Huang, Jun Zhou, Tuan Guo and Bai-Ou Guan. 2017. Real-time, in-situ analysis of silver ions using nucleic acid probes modified silica microfiber interferometry. Talanta 165: 245–250. doi:https://doi.org/10.1016/j.talanta.2016.12.053.

Yuan, Huizhen, Wei Ji, Shuwen Chu, Qiang Liu, Siyu Qian, Jianye Guang, Jiabin Wang, et al. 2019a. Mercaptopyridine-functionalized gold nanoparticles for fiber-optic surface plasmon resonance Hg^{2+} sensing. ACS Sensors 4(3): 704–710. American Chemical Society. doi:10.1021/acssensors.8b01558.

Yuan, Ying-Xue, Jin-Hua Wang and Yan-Song Zheng. 2019b. Selective fluorescence turn-on sensing of phosphate anion in water by tetraphenylethylene dimethylformamidine. Chemistry—An Asian Journal 14(6): 760–764. John Wiley & Sons, Ltd. doi:https://doi.org/10.1002/asia.201801585.

Zeng, Xianzhong and Wei Qin. 2017. A solid-contact potassium-selective electrode with MoO_2 Microspheres as ion-to-electron transducer. Analytica Chimica Acta 982: 72–77. doi:https://doi.org/10.1016/j.aca.2017.05.032.

Zhang, Baozhu and Chunying Wei. 2018. Highly sensitive and selective detection of Pb^{2+} using a turn-on fluorescent aptamer DNA silver nanoclusters sensor. Talanta 182: 125–130. doi:https://doi.org/10.1016/j.talanta.2018.01.061.

Zhang, Ya-nan, Lebin Zhang, Bo Han, Peng Gao, Qilu Wu and Aozhuo Zhang. 2018. Reflective mercury ion and temperature sensor based on a functionalized no-core fiber combined with a fiber bragg grating. Sensors and Actuators B: Chemical 272: 331–339. doi:https://doi.org/10.1016/j.snb.2018.05.168.

Zhang, Junrui, Maneesha Rupakula, Francesco Bellando, Erick Garcia Cordero, Johan Longo, Fabien Wildhaber, Guillaume Herment, et al. 2019. Sweat biomarker sensor incorporating picowatt, three-dimensional extended metal gate ion sensitive field effect transistors. ACS Sensors 4(8): 2039–2047. American Chemical Society. doi:10.1021/acssensors.9b00597.

Zhao, Lijun, Ying Jiang, Huan Wei, Yanan Jiang, Wenjie Ma, Wei Zheng, An-Min Cao, et al. 2019. In vivo measurement of calcium ion with solid-state ion-selective electrode by using shelled hollow carbon nanospheres as a transducing layer. Analytical Chemistry 91(7): 4421–4428. American Chemical Society. doi:10.1021/acs.analchem.8b04944.

Zhao, Long, Zhanlin Zhang, Yuan Liu, Jiaojun Wei, Qingjie Liu, Pan Ran and Xiaohong Li. 2020. Fibrous strips decorated with cleavable aggregation-induced emission probes for visual detection of Hg^{2+}. Journal of Hazardous Materials 385: 121556. doi:https://doi.org/10.1016/j.jhazmat.2019.121556.

Zhong, Nianbing, Zhengkun Wang, Ming Chen, Xin Xin, Ruohua Wu, Yanyan Cen and Yishan Li. 2018. Three-layer-structure polymer optical fiber with a rough inter-layer surface as a highly sensitive evanescent wave sensor. Sensors and Actuators B: Chemical 254: 133–42. doi:https://doi.org/10.1016/j.snb.2017.07.032.

Zhou, Minjuan, Jingjing Guo and Changxi Yang. 2018. Ratiometric fluorescence sensor for Fe^{3+} ions detection based on quantum dot-doped hydrogel optical fiber. Sensors and Actuators B: Chemical 264: 52–58. doi:https://doi.org/10.1016/j.snb.2018.02.119.

Zhou, Junrui, Bowei Li, Anjin Qi, Yajun Shi, Ji Qi, Huizhong Xu and Lingxin Chen. 2020. ZnSe quantum dot based ion imprinting technology for fluorescence detecting cadmium and lead ions on a three-dimensional rotary paper-based microfluidic chip. Sensors and Actuators B: Chemical 305: 127462. doi:https://doi.org/10.1016/j.snb.2019.127462.

Chapter 7

Humidity Sensor

7.1 INTRODUCTION

Humidity is the water vapor content present in the air or other gases. The trace amount of water present can drastically change the physical, chemical, electrical, and mechanical properties of the material. Therefore, it is equally important to detect or monitor the humidity for diverse applications (Blank et al. 2016; Sikarwar and Yadav 2015). Generally, in chemical sensing, a signal is generated when the sensing element interacts with the analyte and is further transduced via a transducer (Wang and Wolfbeis 2020). Therefore, an analyte device with no recognition element is not considered a chemical sensor. So, undoubtedly, humidity sensing is included as a part of the chemical sensor. As humidity is a measure of water in a gaseous state present in the environment and so it follows Dalton's law (Yeo et al. 2008). The humidity is measured in relative humidity (RH) as a proportion of the amount of water vapor present in the atmosphere to the saturated level at the same temperature and pressure. The other term 'moisture' is different from humidity and it is referred to as the water content in liquid form and it only applies to solids. The specific properties of a water molecule are mentioned in Table 7.1.

Numerous conventional methods such as Karl Fischer method and gas chromatography have been used for humidity sensing. However, exhibits certain limitations like slow response time, needs experts, and less portability (Blank et al. 2016; Yeo et al. 2008). There are various other approaches that can fulfill these limitations and show very promising results for humidity sensing. These techniques are based on optical, resistive, capacitance, and mass, etc. Each technique has their

Table 7.1 The properties of a water molecule

Property of H_2O	Respective value
H-O-H angle	104.5°
OH-bond length	0.957 Å
Molar weight	18.01528 g/mol
Molecule radius	1.45 Å
OH-dissociation energy	498 kJ/mol (5.18 eV)
Dipole-moment	6.1×10^{-30}A-s-m (1.83-D)
Acidity (pKa)	13.995
Refractive index(nD)	1.3330 (20 °C)
Thermal conductivity	0.6065 W/m·K

its advantages and limitations, and therefore, the proper selection of technique is needed for better sensing performance.

This chapter reviews the most widely adopted techniques for humidity sensing. The focus is on material choice, design parameters, and sensor characteristics. The possible future scope is also deduced.

7.2 ELECTROCHEMICAL-BASED HUMIDITY SENSORS

A double electrochemical cell containing a solid electrolyte with oxygen ion conductance and another cell consisting of a solid electrolyte with hydrogen ion conductance was developed for humidity sensing (Kalyakin et al. 2020). The humidity analysis in different gas environments (N_2, Ar, and He) under various applied temperatures was analyzed. It was observed that as the temperature increases the corresponding limiting current also increases. This indicated that the diffusion coefficient increased with the rise in temperature and the gas stream from the analyzed channel to the sensor space increased. This study has put a new insight into the effect of temperature on the sensing performance but the humidity range was limited. The graphene-based paper-like structures are attracting a lot of attention due to their simple and easy fabrication, high flexibility, and low weight. Through the proper reduction methods, the GO paper can change to rGO to enhance its electrical conductivity and electrochemical activity (Aksu et al. 2022). Recently, a Janus GO/rGO paper was used as a humidity actuator and with reference to interaction with water, the mechanical movement of the sheet was controlled. This paper contained one side of GO and the other part had rGO structure as shown in Figure 7.1(a,b). The sensing mechanism was based on the folding of Janus GO/rGO paper with the adsorption of moisture. It was proposed that due to the hydrophobic nature of the rGO surface, it repelled the water molecules and to the other side, the hydrophilic GO swelled with respect to the adsorption of water molecules. This led to producing stress stress from the GO side to the rGO surface and resulted in a circular folding. When there will no moisture, the folding structure will get back to its normal position. It was observed that the flexibility decreased for thicker paper structures so it is very important to control the paper thickness to obtain good elasticity and also the rise and decay time of the device.

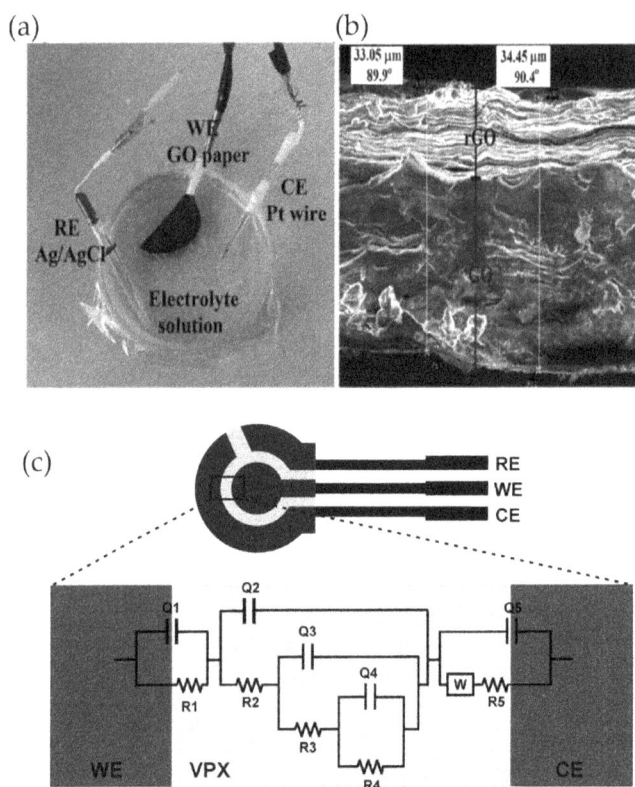

Figure 7.1 (a, b) The fabrication of Janus GO/rGO paper-based humidity sensor. [Reproduced with permission (Aksu et al. 2022). Copyright 2022, Elsevier]. (c) Vanadium pentoxide xerogel modified screen–printed graphite used humidity sensor and its equivalent electric circuit. [Reproduced with permission (Trachioti and Prodromidis 2020). Copyright 2020, Elsevier].

The metal oxides are widely used for the detection of humidity sensing due to their physicochemical properties. When water molecules get adsorbed at the oxide–air interface, it forms a layer-by-layer formation on the material surface. This tends to alter the electrical resistance of the material and hence, the conductivity of the material. The change in electrical resistance is proportional to the adsorption rate of the humidity.

For example, the metal oxides $V_2O_5 \cdot nH_2O$ based electrochemical sensor explored for the determination of humidity (Trachioti and Prodromidis 2020). This material had a great affinity to water molecules. Additionally, the $V_2O_5 \cdot nH_2O$ structure contained double layers of square pyramidal VO_5 parts, disconnected by water molecules, and exhibited 100 times greater electrical conductance related to anhydrous orthorhombic V_2O_5. Here, the TiO_2:WO_3 pair doped with three different quantities of V_2O_5 were fabricated. The equivalent electric circuit of the sensor design is shown in Figure 7.1(c). This sensor showed a large sensitivity of 190–500 Ohm/RH% with a rise and decay time of 52 s/21 s to 10–93% RH being observed. Another metal doped metal oxide material was also utilized for the detection of humidity sensing. In short, the Al^{3+} doped SnO_2 nano powders were synthesized by hydrothermal method, and among that the $Sn_{1-x}Al_xO_2$ (x = 0.05) showed good electrochemical properties (Blessi et al. 2021). The main reason for choosing the metal was that as the substitution of Al^{3+} by Sn^{4+} ions in the SnO_2 matrix created oxygen vacancies due to charge imbalance and generated extra free electrons in the conduction band. The doping increased the surface area and porosity which led to good response and wide bandwidth. However, the temporal characteristic was in minutes and further needs to be decreased. The MOFs are also extensively used for humidity sensing due to maximum adsorption area, more pores, and being thermally stable (Sunilkumar et al. 2019). The reported structure exhibited good hydrophilicity and conductivity. The sensor showed a response around 10^3 to 23–95% RH. Over for 30 days, the sensor showed some fluctuations in the sensing response and also the repeatability test of 9 cycles displayed the decrease in current. Alternative polypyrrole-tantalum disulfide (PPy-TaS_2) composites were tested for humidity sensing. In this work, the wt% of TaS_2 (10, 30 and 50) in PPy were varied and noticed that the PPy/TaS_2-50% showed its best humidity sensing performance (Sunilkumar et al. 2019). The TaS_2 was used due to its large affinity of Ta in TaS_2 and hence, encouraged more energy for the adsorption of moisture. The separation between the layers also provided a surface charge thus, enhancing the response. As compared with earlier reported work, the long-term stability, response and recovery times were quite improved. The impedance-type humidity sensing was also performed by using bare 2D materials. These types of materials showed specific properties like a large-area, uniform single-layer and high carrier density. The sensor showed a high response of more than 10^4 from 0% to 35% RH (Zhao et al. 2017b). Despite its high sensitivity, the device recovery time was quite high, and also the humidity bandwidth was smaller.

The inorganic halide perovskites have also been explored for humidity sensing owing to their interesting features like good chemical and thermal stability. Among perovskites materials, the $CsPbBr_3$ nanocrystals are widely used perovskites for humidity sensing due to their high air-stable characteristics, fewer surface defects and better stability, and good surface charge carriers (Wu et al. 2021). Recently, it was used for humidity sensing by using an impedance method. Even at low working voltage (20 mV), the device displayed a sensitivity of 1.5565%/RH% and fast time-response characteristics of just 2–3 s to humidity at room temperature. The CNTs showed promising results for humidity sensing due to their excellent electro-optic, physical properties, flexibility, and large surface area. Recently, cellulose nanofiber (CNF) combined with CNTs-based paper structure was utilized for the wide detection of humidity range from 11–95% RH (Zhu et al. 2020). The sensor displayed a sensitivity of 65.0% ($\Delta I/I0$) to 95% RH. But sensor response and recovery times were very high at 321 s/435 s, respectively. A summary of a few electrochemical-based humidity sensors is shown in Table 7.2.

Table 7.2 Some electrochemical-based humidity sensors

Materials used	Detection range (% RH)	Response/ recovery time	References
$CaZr_{0.9}Sc_{0.1}O_{3-\delta}$	0.8–2.7	Initial response time: 50 s	(Kalyakin et al. 2020)
GO/rGO paper	30–80	68 s/53 s	(Aksu et al. 2022)
GO/rGO paper	30–80	68 s/53 s	(Aksu et al. 2022)
Al^{3+} substituted SnO_2 nanopowder	10–95	87 s/64 s	(Blessi et al. 2021)
PS/PPy	11–97	54.9 s /76.8 s	(Aguiar et al. 2021)
PPy/WS_2	11–97	52 s/58 s	(Sunilkumar et al. 2019)
Sc@calcium zirconate ($CaZr_{0.95}Sc_{0.05}O_{3-\delta}$; CZS), and Sc-doped barium stannate ($BaSn_{0.75}Sc_{0.25}O_{3-\delta}$, BSS)	In nitrogen (pH_2O = 0.025–0.049 atm) for CZS electrolyte and in nitrogen atmosphere (pH_2O = 0.025–0.049 atm) for BSS	–	(Kalyakin et al. 2021)
PPy/TaS_2	11–97	10 s/ 20 s	(Sunilkumar et al. 2019)
Co-MOF@PA	23–95	179 s/27 s @54% RH	(Huo et al. 2022)
$FeCl_3$-NH_2--MIL-125(Ti)	11–95	11 s/86 s	(Zhang et al. 2014)
MIL-101(Cr)	33–95	17 s/90 s	(Zhang et al. 2017)
KOH/M050	20–90	36 s/118 s	(Su and Lee 2018)
MoS_2	0–35	10 s/60 s	(Zhao et al. 2017b)
$CsPbBr_3$ nanoparticles	11–95	2.8 s/9.7 s	(Wu et al. 2021)
CNF/CNT-coated paper	11–95	321 s/435 s	(Zhu et al. 2020)
MWCNTs-coated paper	11–95	470 s/500 s	(Zhao et al. 2017a)

7.3 CAPACITIVE-BASED HUMIDITY SENSORS

Several capacitive-based sensors have been extensively tested for humidity sensing applications as these kinds of devices convert humidity change into their respective capacitance. This has specific advantages such as simple structure, low-cost, less power consumption and hysteresis, and linearity. With respect to changes on the humidity exposure, the dielectric constant of the dielectric material changes and thus, modulates its capacitance. A summary of a few capacitive-based humidity sensors is shown in Table 7.3.

Table 7.3 Summary of a few capacitive-based humidity sensors

Materials used	Deposition method	Detection range (%RH)	Sensor sensitivity	Response/recovery time	References
Mg$_2$(dobdc)	Spraying method	0–97	–	2.5 s/1.5 s	(Zhang et al. 2022a)
BaTiO$_3$	Inkjet printing	20–80	5.75×10^5 pF/% RH	41 s/34 s	(Fernandez et al. 2021)
Poly (vinylidenefluoride-co-trifluoroethylene) [P(VDF-TrFE)]	Spin-coating	50–90	–	3.693 s/3.430 s	(Niu et al. 2021)
MgAl$_2$O$_4$	Screen printing	2–98	~109 pF at 89% RH	~66 s/~71 s	(Das et al. 2021)
1050 Al alloy	Anodization process	20–80	984.2 nF at RH 80%	9 s/11 s	(Chung et al. 2021)
CuCr$_2$O$_4$ nanopowders	Screen printing	1–98	3.86 nF to 4.5 nF with 40%–80% RH	3.6 s/~128 s	(Mahapatra et al. 2021)
Pyromellitic dianhydride (PMDA)/ p-phenylenediamine (p-PDA), and PMDA/oxydianiline (ODA)-TiO$_2$	Spin-coating	10–90	0.94 pF/% RH	40 s	(Yu et al. 2021c)
PEDOT:PSS	Spin-coating	52.0–93.4	0.034 pF/% RH	30 s/1 min	(Yao and Cui 2020)
PVDF-BaTiO$_3$	Spin-coating	40–90	0.2416 pF/% RH	40 s/25 s	(Mallick et al. 2020)
PMCM-41/PEDOT	–	10–90	–	165 s/115 s	(Qi et al. 2020)
SnO$_2$/MoS$_2$	Drop casting	45–90	3170 pF/% RH	17 s/6 s	(Zhao et al. 2018)
BPDA/6-FDA	Spin-coating	30–90	~6.0 pF	–	(Kim et al. 2020)
GO	Immersion method	30–90	209% (ΔC/C0) at 1 kHz	–	(Alrammouz et al. 2019)

(Contd.)

Table 7.3 Summary of a few capacitive-based humidity sensors (*Contd.*)

Materials used	Deposition method	Detection range (%RH)	Sensor sensitivity	Response/ recovery time	References
ZnO nanorods/WS$_2$ nanosheets	Spin-coating	18–85	101.71 fF/% RH	74.51 s/ 25.67 s	(Dwiputra et al. 2020)
TiO$_2$ nanoparticles	Spin-coating	10–90	1.24 pF/% RH	25 s	(Qiang et al. 2018)
PANI decorated Cu–ZnS porous microspheres	Spin-coating	30–90	12 pF/RH	42 s/24 s	(Parangusan et al. 2020)
Keratin (This work)	Drop casing	16–92	0.16 pF/% RH	41 s/62 s	(Hammouche et al. 2021)
Keratin/1%GO	Drop casing	16–92	12.27 pF/% RH	39 s/80 s	
Keratin/1%CF	Drop casing	16–92	633.12 pF/% RH	21 s/56 s	
Collagen	Drop casing	50–90	1.14 fF/% RH	–	(Vivekananthan et al. 2018)
Cellulose	Writing graphite pencil on paper	30–90	–	10 s/32 s	(Kanaparthi 2017)

Recently, MOFs have been used for humidity sensing because of their straight contact among ion transporters and water vapors, high response, fast sensing speed, and environmental stability. In detail, the MOFs consisting of several metallic ions acted as charge transporters rather than electrons (Zhang et al. 2022a). Here, different metallic ions like magnesium, copper, zinc, cobalt, and nickel along with various organic ligands of terephthalic acid (H4PTA), 1,4-dioxido-2,5-benzenedicarboxylate (H4dobdc), and pyromellitic acid (H4PMA) have been used and optimized. This sensor showed a wide detection range of 0% to 97% RH with fast response characteristics of just 2.5 s and 1.5 s, respectively. The sensor's cross-selectivity to other interfering liquids was not tested. Nowadays, various ceramics materials have been used in humidity sensing applications due to their numerous advantages like high stability, large working temperature, heat resistant, and sensitivity to various analytes. Recently, ceramics materials such as barium titanate-based capacitive humidity sensor have been reported. This material exhibited properties like high hydrophilic and large dielectric nature. This sensor displayed a response ~575000 pF/%RH, hysteresis 8.9%, rise and decay time ~41 s, and ~30 s, correspondingly (Fernandez et al. 2021). When water adsorbed onto the sensing layer, the protonic charge process increased and thus, increased capacitive detection ability. This study did not report the cross-selectivity effect on humidity performance.

A transition metal-based spinel structure of Copper Chromite (CuCr$_2$O$_4$) was also used (shown in Figure 7.2(a)) for the detection of humidity (Mahapatra et al. 2021). The high defect density of this material can be used for the monitoring of humidity. Over the humidity range of 0–40% RH, the total change in the capacitance of the sensor from 3.86 nF to 4.5 nF was reported. The sensor geometry was prepared by a screen-printing technique. The sensor response long-term stability was incredible at around 6 months. However, the recovery time of the sensor was longer ~128 s than the response time of just ~3.6 s only. A simple drop cast using a capacitive sensor was fabricated for humidity sensing. The nanocomposites of SnO$_2$-modified MoS was synthesized by a hydrothermal route and successfully employed for humidity sensing (Zhao et al. 2018). The schematic of the humidity sensor is shown in Figure 7.2(b). The 2D materials like MoS$_2$ have a high surface-to-volume ratio, good hygroscopicity, conductivity, and high carrier mobility. To improve the sensor's functioning, it can be clubbed together with metal oxides like SnO$_2$. Due to heterojunction properties, the sensor was able to show a very high response of 3170 pF/% RH to 45–90% RH detection range with fast switching time. In the high detection range, the response of the sensor followed an exponential relationship. The sensor long-term stability and selectivity to moisture only needs to be analyzed.

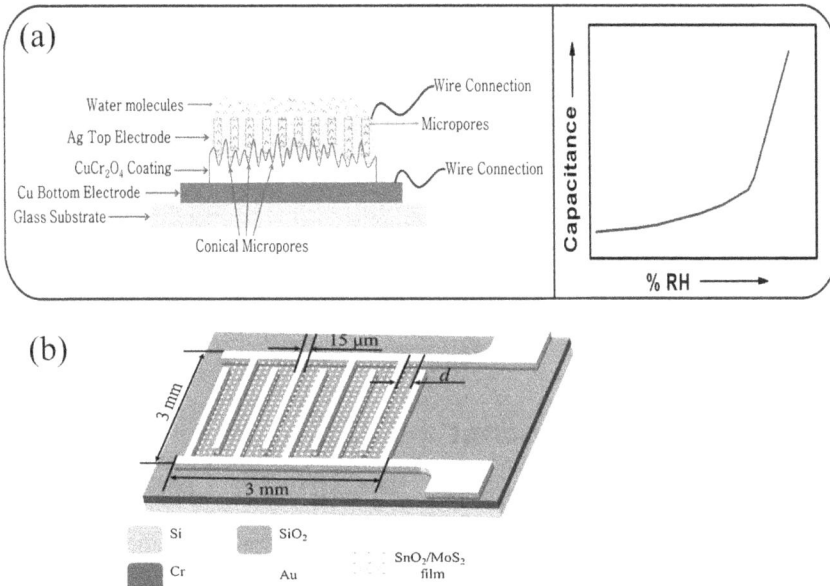

Figure 7.2 (a) Copper chromite thick film based novel and ultrasensitive capacitive humidity sensor. [Reproduced with permission (Mahapatra et al. 2021). Copyright 2021, Elsevier]. (b) Schematic of the SnO$_2$/MoS$_2$ capacitive humidity sensor. [Reproduced with permission (Zhao et al. 2018). Copyright 2018, Elsevier].

7.4 RESISTIVE-BASED HUMIDITY SENSORS

The resistance-based sensors are extensively used for humidity detection. Generally, the materials resistance changes when it is exposed to a humid environment. The advantages of such method include a simple structure, fast response, long-term stability, and high mechanical stability. Figure 7.3 shows some sensors development for humidity sensing. The recent development in resistive-based humidity sensors is shown in Table 7.4.

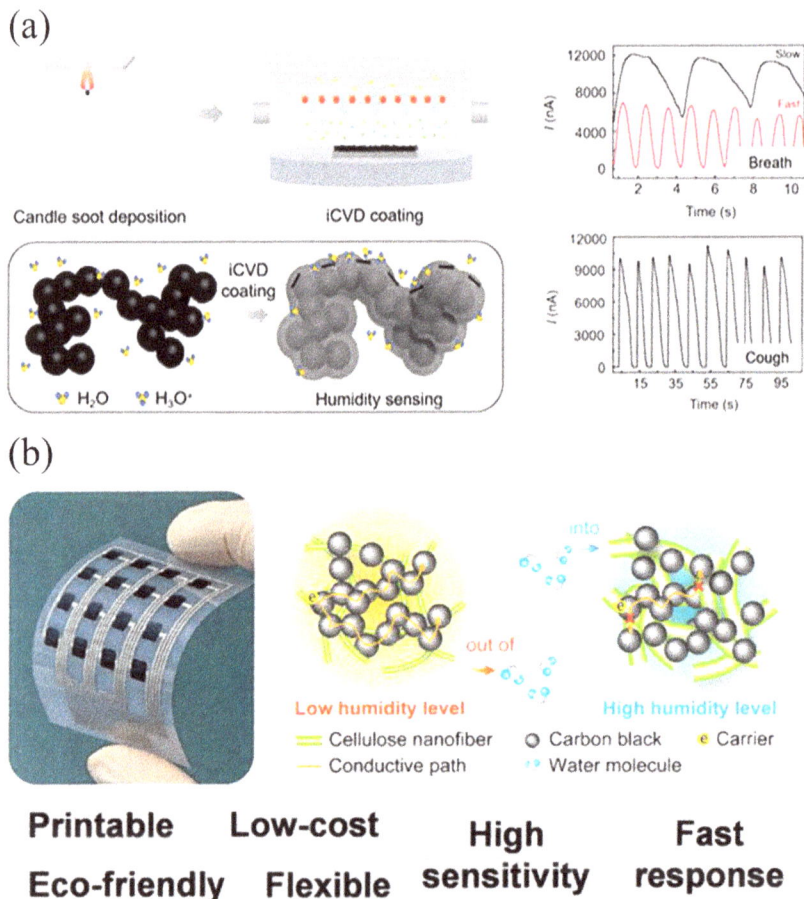

Figure 7.3   (a) Schematic of candle soot deposition on a glass slide and iCVD coating of p(DMAEMA-co-EGDMA) on CS film and its utilization for humidity sensing. [Reproduced with permission (Su et al. 2022). Copyright 2022, Elsevier]; (b) Representation of humidity sensor mechanism based on CNF/CB composite. [Reproduced with permission (Tachibana et al. 2022). Copyright 2022, American Chemical Society].

Table 7.4 A summary of some resistive-based humidity sensors.

Materials used	Detection range (%RH)	Sensor sensitivity	Response/ recovery time	References
ZnO-cellulose	40–90	4.487 MΩ/% RH	8 s/10 s	(Sahoo et al. 2020)
PPy/ZnO	5–95	0.31/% RH	12 s/ 8 s	(Shukla et al. 2018)
Graphene-coated cellulosic paper	4–89	3.5 kΩ/% RH	9 s/15 s	(Khalifa et al. 2020)
Graphene/Polymer	0–97	20000@97% RH	20 ms/17 ms	(He et al. 2018)
PVA/graphene	10–80	66.4%@80% RH	11 s/35 s	(Chen et al. 2021a)
Graphene oxide (GO)/poly (3, 4-ethylenedioxythiophene)/ poly(styrenesulfonate) (PEDOT: PSS)/Ag colloids (AC)	12–97	3.5% @97% RH	54 s/132 s	(Pang et al. 2018)
WS_2	20–90	2357@90% RH	5 s/6 s	(Guo et al. 2017)
Si nanocrystals	8–83	–	40 ms/40 ms	(Kano et al. 2017)
SiO_2 NPs	10–93	10000@93% RH	31.4 s/6.5 s	(Kano and Fujii 2018)
Cellulose/KOH	11.3–97.3	–	6.0 s/10.8 s	(Wang et al. 2020)
Carbon nanocoils	4–95	12% @80% RH	1.9 s/1.5 s	(Wu et al. 2019)
MWCNTs/PLL	0–91.5	6.6@91.5% RH	30 s/2 s	(Zhao et al. 2019)
TaS_2 nanosheets	11–95	187.6@95% RH	0.6 s/2.0 s	(Feng et al. 2019)
Ag nano-lines (AgNLs)	11–94	Resistance change: ~87%	< 0.9 s/< 3.1 s	(He et al. 2022)

The cellulose and its derivatives exhibit volume change when it interacts with the water molecules and so, widely investigated for humidity sensing applications. Recently, a ZnO nanocrystal grown on cellulose fibers via an aqueous chemical bath deposition method was investigated for humidity sensing (Sahoo et al. 2020). The ZnO was used because of its tunability, large surface area, and good binding with cellulose fibers material. On another side, cellulose was hydrophilic in nature and so it becomes a porous nanocomposite structure thus, water molecules can easily be adsorbed on the surface of the nanocomposite and therefore, very effective for humidity sensing. This feature allowed this sensor to display a high sensitivity of 4.487 MΩ/%RH in the humidity range of 40–90%. In the repeatability test, only two cycles were considered and also, the sensor performance in low humidity exposure was not studied. The metal oxides and polymer nanocomposites were used for the detection of humidity. In detail, the weight percentage of zinc oxide in polypyrole was optimized and explored for the detection of humidity in the range 5–95% RH (Shukla et al. 2018). The sensing mechanism was based on the adsorption, decrease in resistance, and ionizability of the hybrid composite. The effect of interferents like ethanol, acetone, and ammonia with water adsorption was considered, and observed the negligible effect on the sensing performance. There was a little deviation of resistance over the humidity range for the composite sample. However, the sensor bandwidth was also very high. A graphene coated on cellulose paper via dip-coating method and vacuum filtration method was fabricated (Khalifa et al. 2020). Graphene is promising for humidity sensing applications due to its high specific area, excellent electron mobility and ability to work in a wide temperature range. The addition of cellulose in graphene contributed to the improvement in the sensing performance due to high surface area, porosity, flexibility, low weight, and high electron transport. The detection range was increased as compared to metal oxides added to cellulose. However, the temperature effect needs to be considered in the cellulose properties. The GO, poly (3,4-ethylenedioxythiophene)-poly (styrenesulfonate) (PEDOT: PSS) and Ag colloids (AC) are employed for the detection of humidity (Pang et al. 2018). Due to the porous structure, it exhibited a fast response and recovery time. This study did not report long-term stability.

A flexible and highly sensitive 2D large-area WS$_2$ film for humidity sensing in naturally flat and high mechanical flexible states was projected (Guo et al. 2017). The sensing geometry contained graphene as an electrode, thin polydimethylsiloxane (PDMS) as a base and WS$_2$ as a sensing element. The sensor was studied under stressed, compressed, and relaxed states. This sensor was easily laminated onto human skin and displayed a good sensing response to humidity under motion. Though, the cross-selectivity to other interfering agents and long-term stability study over few days was not considered. Recently, an interdigital electrode based Ag nano-lines (AgNLs) coated sensory platform was tested for humidity sensing (He et al. 2022). The AgNLs was built by the electrospinning and UV irradiation reduction of PVP nanofibers. When this structure was exposed to water molecules, it formed a conductive path across these solver lines, which was detected by measuring the variation in resistance. While considering this type of platform, the uniform mixing of polymer and metal nanoparticles must be considered.

7.5 OPTICAL FIBER HUMIDITY SENSORS

In most the optical and electronic interfaces, the light is guided by an optical fiber, and therefore, vastly used due to its fascinating properties like small size, high sensitivity, immunity to electromagnetic interference, chemical inertness, multiplexing capability, and robustness to harsh environments. A few optical fiber geometries explored for humidity sensing are shown in Figure 7.4. Table 7.5 shows a summary of a few optical fiber humidity sensors.

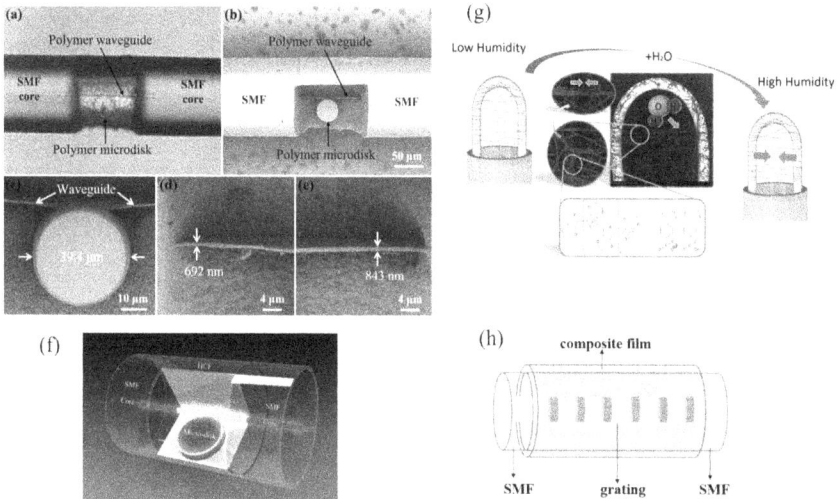

Figure 7.4 (a-f) Planar view of the fs laser-printed microdisk WGM resonator including a polymer waveguide and a polymer microdisk in a SMF. (a) Microscopic image. (b) SEM of complete resonator. (c) microdisk and (d) waveguide left side, (e) right side, and (f) Graphic of in-fiber polymer microdisk WGM resonator. [Reproduced with permission (Ji et al. 2021b). Copyright 2021, American Chemical Society]; (g) Diagram of the hydrogen-bonding interface used in the tennis racket-shaped fiber made with a polymer solution having PVA@PEDOT:PSS. [Reproduced with permission (Wen et al. 2020). Copyright 2020, Elsevier]; (h) Long-period fiber grating (LPFG) coated PEG/PVA film for humidity sensing. [Reproduced with permission (Wang et al. 2019a). Copyright 2019, MDPI].

Polymers are very effective for humidity sensing. When water molecules interact with the polymeric materials, it starts swelling which corresponds to increase in length of the interlayer distance and changes the refractive index and thus, modulates the reflectance. The

Table 7.5 Summary of optical fiber-based humidity sensors

Optical fiber type	Materials used	Detection range (%RH)	Sensor sensitivity	Response/recovery time	References
SMF-FC-SMF	Polyvinyl alcohol (PVA)	55–85	1.78 nm/% RH	12 s	(Yu et al. 2021b)
FP cavity	PVA	45–74	0.248 nm/% RH	–	(Chen et al. 2021b)
Double D-shaped optical fiber	PVA	20–80	11.6 nm/% RH	–	(Wang et al. 2021)
Hollow-core fiber (HCF)/chitosan cavity/air-chitosan hybrid cavity	Chitosan film	40–92	7.15 nm/% RH	0.98 s/0.68 s	(Zhou et al. 2021)
FP based	Chitosan	30–95	0.081 nm/% RH	0.08 s/0.07 s	(Shrivastav et al. 2020)
Long-period fiber grating (LPFG)	Polyethylene glycol (PEG)/polyvinyl alcohol (PVA) composite	50–75	2.485 nm/% RH	–	(Wang et al. 2019a)
LPFG based	Graphene oxide (GO)	20–80	−0.1824 dB/% RH	–	(Tsai et al. 2021)
Modal Interferometric	SnO_2	20–90	0.31 nm/% RH	–	(Lopez-Torres et al. 2017)
Polymer cap onto optical fiber facet	NOA 81	10–95	148 pm per % RH	–	(Arrizabalaga et al. 2019)
FPI based	HA/PVA	35–85	−525 pm/% RH	2 s	(Zhang et al. 2022c)
SPR based	GQDs-PVA	22.97–85.46	23 pm/% RH	–	(Tong et al. 2022)
Hybrid functional tip	Poly(allylamine hydrochloride) (PAH)/SiO_2 NPs	55–90	0.43%/% RH	3.1 s	(He et al. 2021)
FPI based	Thymol blue in porous Ormosil film	0–90	~0.19 nm/% RH	32 s/56 s	(Liu et al. 2022)
Two single-mode fibers (SMFs) spliced on two sides of a no core fiber (NCF)	Calcium alginate (CaAlg) hydrogel	30–80	0.3774dBm/% RH	3 s/4 s	(Bian et al. 2020)
Tennis racket-shaped	PVA@PEDOT:PSS	20–80	−0.990 nm/% RH	–	(Wen et al. 2020)
Fiber Bragg grating (FBG)/hollow-core fiber (HCF)	Carboxymethyl cellulose (CMC)/carbon nanotubes (CNTs)	35–85	230.95 pm/% RH	–	(Li et al. 2020)
Fiber Bragg grating (FBG)	Polyimide (PI)	11–83	2 pm/% RH	–	(Zhang et al. 2021)
Balloon shaped interferometer	Au NPs	35–95	−0.571 nm/% RH	0.00133 s/ 0.00186 s	(Al-Hayali et al. 2021)
LPGs based	Au/SiO_2	30–90	0.53 nm/% RH	–	(Hromadka et al. 2019)

fiber optics manipulated with polymer was extensively studied for humidity sensing. For example, polyvinyl alcohol as a sensing material was used for humidity sensing due to its fascinating sensing of thin-film construction ability. The sensor was fabricated by coating the PVA material which formed an FPI cavity (Yu et al. 2021b). The FPI sensor showed a sensitivity of 1.7 nm/%RH. The sensing mechanism is dependent upon the vernier mechanism in which the dual spectrum superimposing principle was used. However, it took almost 1 h to get it stabilized. The PVA might be affected by the temperature variation. Another polymer composite coated on fiber grating type of LPGs was used to detect humidity (Wang et al. 2019a). The polyethylene glycol (PEG)/polyvinyl alcohol (PVA) mixed film coated onto the LPGs, which was sensitive to change in humidity in terms of its change in effective refractive index and thus, changed the resonance dip of the fiber grating. The change in the resonance dip was due to a change in the refractive index of the materials only. Also, the polymer composite refractive index was nearer to fiber cladding and so, very sensitive to humidity variation. The proposed design was sensitive to 50–75% RH with obtained sensitivity of 2.485 nm/%RH. This study did not report the rise and decay time of the device. Metal oxides were also utilized in the fiber optics domain to detect the humidity. The SnO_2 nanofilm was sputtered onto the PCF, which was spliced between the two SMFs to excite the cladding modes into the fiber (Lopez-Torres et al. 2017). The thickness of the film varied over 470–1800 nm and at the optimized thickness of 1150 nm, the sensor showed a large response of 67 nm to humidity change from 20 to 90% RH. The study indicated that the device exhibited good linearity of up to 80% RH. Recently, an SPR-based humidity sensor using humidity sensitive materials like graphene quantum dots and polyvinyl alcohol (GQDs-PVA) was reported (Tong et al. 2022). The PCF spliced between two MMFs along with the material coating overlayed onto the PCF and the corresponding transmission spectrum under humidity variation was recorded. The sensor exhibited a response of 23 pm/%RH. Although, the sensor's long-term stability was limited up to 10 hours only.

A tennis racket like optical fiber coated with polymeric liquid containing poly (3,4-ethylenedioxythiophene) polystyrene sulfonate (PEDOT:PSS) fixed in polyvinyl alcohol (PVA) by an electrospinning method was utilized for humidity sensing (Wen et al. 2020). The schematic of an optical fiber is displayed in Figure 7.4(g). The light followed via whispering gallery mode (WGM) theory. Under humidity variation, the refractive index of the sensing material changed and hence, corresponding light interaction altered which was reflected in the transmission spectrum. The sensor displayed a response of −0.990 nm/%RH during 20 to 80% RH. A fiber grating was also used in humidity sensing due to its sensitive nature with respect to a small changes in the refractive index of the sensing material. The polyimide (PI) coated fiber Bragg grating by a dip coating method (Zhang et al. 2021). A temperature compensation method by monitoring the Bragg wavelength with temperature change was used. The sensor showed a broad detection range of 11% RH to 83% RH, with a sensitivity of 2 pm/% RH. However, as compared to other sensors, the sensitivity is limited and needs to be handled properly due to the delicate nature of FBGs.

7.6 LUMINESCENT AND FLUORESCENT BASED HUMIDITY SENSORS

The luminescence-based sensing approach has been considered for humidity sensing due to its fast response, high sensitivity, and simple design structure, low-cost setup. The following are some works based on this technique for humidity sensing. A summary of a few luminescent-based humidity sensors is shown in Table 7.6.

Numerous MOFs have been extensively used for humidity sensing because of tunability in structural and luminescent properties. In detail, the Ba-MOFs and Sr-MOFs were used for the detection of humidity at 30, 40, 50, and 85% RH (Fard et al. 2016). In the study, it was observed that the color of these materials changed from white to bright yellow. The corresponding wavelength exhibited a red-shift from 505 nm and 522 nm. The materials showed a red and blue-shift after exposure to humidity of 100% and de-exposing. The cross-interference from other vapors was not discussed. The perovskites material has also attracted attention towards its use for humidity sensing due to its large ionic nature and aqueous solubility. Among the various perovskite's materials, the $CsPbBr_3$ perovskite is very sensitive to humid environments. By using the $CsPbBr_3$ perovskite-covered paper substrate-based humidity sensor was realized (Xiang et al. 2021). It was reported an evaporative sensor for water content. The linear relation of decrease in intensity with respect to increasing in humidity level was obtained. The material long-term stability was not discussed.

Recently, vapoluminescence-based platinum salt was used for humidity detection and obtained a high sensitivity and stability. The platinum (II) salt of crystal and powder forms under different humidity ranges were tested (Norton et al. 2022). When it was exposed to water vapor, the water molecules got incorporated in the crystal lattice and thus reflected in terms of changes in color from dark red and luminescence to red (Figure 7.5). In presence of interfering species, this material showed good selectivity and stability.

Table 7.6 A summary of a few luminescent-based humidity sensors

Materials used	Sensor sensitivity	Detection range (%RH)	References
Ba-MOFs, Sr-MOFs	–	30–100	(Fard et al. 2016)
Hyaluronic acid (HA) with lanthanide ions (Tb^{3+} and/or Eu^{3+})	43.15%@98% RH	16–98	(Xia et al. 2021)
$CsPbBr_3$	96.7–102.5%	1–17	(Xiang et al. 2021)
$CsPbBr_3$	7500 a.u. @33%	33–98	(Xiang et al. 2021)
Platinum (II) salt, $[Pt(tpy)Cl]ClO_4$ (tpy = 2,2′;6′,2″-terpyridine)	Intensity (a.u.) = -3×10^7 log (humidity %) + 5×10^7	–80% for crystals; 30–60% for powders	(Norton et al. 2022)
Zeolite/$CsPbBr_3$	1978.83 a.u./RH %	0–92%	(Zhang et al. 2022b)
$CH_3NH_3PbBr_3$	0.1188 a.u./RH %	7–98	(Xu et al. 2016)
Tb^{3+}@p-CDs/MOF)	–	33.0–85.1	(Wu and Yan 2017)
Eu-MOFs	8×10^4 a.u.	0–100	(Gao et al. 2017)

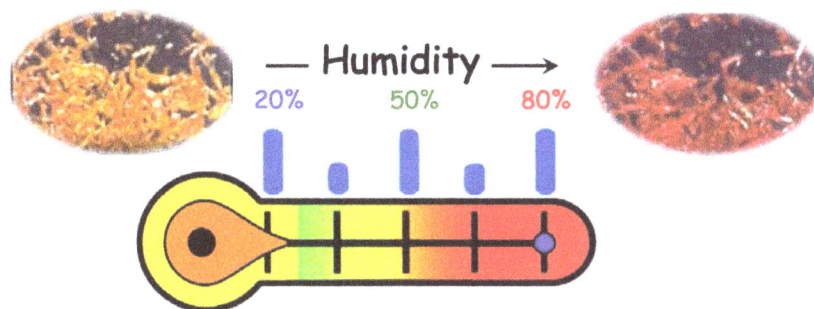

Figure 7.5 Vapoluminescence-based platinum (II) salt-used for humidity sensing. [Reproduced with permission (Norton et al. 2022). Copyright 2022, Elsevier].

7.7 COLORIMETRY-BASED HUMIDITY SENSORS

Certain materials change their color when any analyte comes in contact with the sensing material. Due to this feature, these materials are extensively used in environmental monitoring applications like humidity sensing.

Here, some materials are listed for humdity sensing. Table 7.7 shows a summary of a few colorimetric-based humidity sensors.

The cellulose nanocrystals (CNCs) are considered the best materials for humidity sensing due to their different coloring abilities, high modulus, optical transparency, anisotropy, and flexibility. But due to poor water adsorption capability and to increase the selectivity to humidity, the composite material was used. In detail, the CNC/poly(N-isopropylacrylamide) (PNIPAM) was tested to a humidity range of 70 to 97% RH (Sun et al. 2019). For CNC-30 sample, the reflectance spectrum showed a red shift up to ~100 nm to humidity changed from 70 to 97% RH. It observed that as the water adsorption increased, the visibility decreased and also it took around 150 seconds to come back to its original signal. Another CNC-based colorimetry humidity sensor was also reported. In this, the polyol, i.e., glycerol (G), xylitol (X) and sorbitol (S) was used as a plasticizer which improved the elasticity of CNC film (Meng et al. 2020). This sensing performance of over 30–95% RH was studied. When these samples were exposed to varied humidity, the shed of CNC-G, CNC-X, and CNC-S slowly altered from light blue to orange, light steel blue to pink, and light green to red, respectively. The response time to show the color change was not mentioned. The indicator dyes are also used for humidity sensing due to their ability to color change in presence of humidity. This process is reversible and so widely adopted for humidity detection. Generally, only indicator dyes or inorganic salts cannot be utilized for the sensor preparation. The host material should be used so that the indicator will be dispersed uniformly. For example, a bromothymol blue (BTB) sulfonphthalein dye along with titanium dioxide, palygorskite, and mulite which acted as a host were explored for humidity detection (Wang et al. 2019b). The results showed that the BTB-palygorskite device displayed a fast response and stability under varied humid environments due to appropriate surface area and absorption of liquids. For BTB-palygorskite material, it changed from pinkish to bright yellow under 11, 43, and 75% RH. It was detected that the surface area of the minerals played a crucial role in deciding the response. However, good color stability was observed in 4 minutes, which was quite long for speedy detection requirements. Recently, the carbon materials like GO were used for humidity sensing. The GO was coated on various substrates like silicon, aluminum foil (Al), and metalized PET (mPET) using a dip coating method (Chi et al. 2021). It was observed that the color change property of GO on mPET prepared by aqueous dispersion showed a poor resolution and sensitivity whereas, these properties greatly enhanced when it was prepared from acetone dispersion. Due to wetting property of the solvent, which formed a uniform d-spacing of GO sheet could be the reason behind this improvement. This method was sensitive and did not require any electrical source, was cheaper, and could be easily fabricated. This study did not discuss the response time of the sensor. Due to high surface area and interconnection, pore channels of silica were used for humidity sensing. It was observed that with respect to the exposure of water molecules on mesoporous silica, the phenomenon of capillary condensation was triggered, which was dependent upon the pore size of the silica (Švara Fabjan et al. 2020). The selective functionalization of methylene blue as an indicator dye was also done. The methylene blue was chosen because of its nontoxicity in low concentration. With respect to the increase in humidity from 55, 79, and 88% RH, the methylene blue adsorbed on the surface of mesoporous SiO_2 was dissolved, which led to significant blue coloration occurred. This sensor was simple and low-cost. While the optimization of an indicator dye with respect to pore size needs to be considered.

Table 7.7 Summary of a few colorimetric-based humidity sensors

Materials used	Detection range (%RH)	Response/ recovery time	References
Cellulose-PNIPAM	70–90	30 s/180 s	(Sun et al. 2019)
Cellulose/polyol	30–95	–/–	(Meng et al. 2020)
BTB-palygorskite, BTB-titanium dioxide and BTB-mullite	@11, 43 and 75	1 min	(Wang et al. 2019b)
{[Co$_2$(DPNDI)(2,6-NDC)2]·7(DMF)}n (1)	@45	10 s	(Qin et al. 2019)
Hydrogel	5–95	–	(Yu et al. 2021a)
GO	0–96	–	(Chi et al. 2021)
Retroreflective structural color film (RRSCF)	50–80	–	(Ji et al. 2021a)
Poly(DCDA-HC1) and poly(DCDA-HC2)	@80	–	(Mergu et al. 2020)
Opal photonic hydrogel	25–95	<3 s	(Sobhanimatin et al. 2021)
Poly(acrylamide-N,N′-methylene bis(acrylamide)) (P(AM-MBA)) nanogels and TiO$_2$ nanoparticle	47.0–89.3	0.5 s	(Kou et al. 2018)
PVA/SiO$_2$ opal structure	0–100	<1 s	(Yang et al. 2017)
PAAm–P(St–MMA–AA) PC hydrogel	20–100	–	(Tian et al. 2008)
Poly(2-hydroxyethyl methacrylate) (PHEMA)	0–80	0.6 min /2.8 min	(Kim et al. 2018)
Mesoporous SiO$_2$/dye	53–92	–	(Švara Fabjan et al. 2020)

7.8   QUARTZ CRYSTAL MICROBALANCE-BASED HUMIDITY SENSORS

The quartz crystal microbalance (QCM) is a piezoelectric device that is enormously responsive to minute variations in the mass of the molecules. It is highly dependent upon the materials physico-chemical properties of the detecting material. The following are some QCM-based humidity sensors based on numerous materials. Some quartz crystal microbalance-based humidity sensors are shown in Figure 7.6. Table 7.8 shows some quartz crystal microbalance-based humidity sensors.

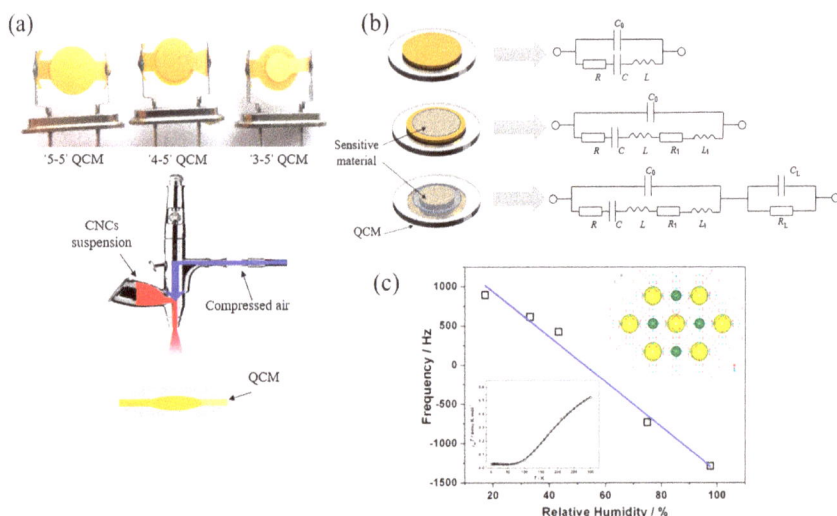

Figure 7.6   (a) Schematic of QCM humidity sensor prepared by using cellulose nanocrystals coated three different electrode structures and its coating process by spraying method, (b) Its equivalent electronic model with reference to with and without coated materials. [Reproduced with permission (Yao et al. 2020). Copyright 2020, Elsevier]. (c) Humidity sensor response characteristics based on copper metal and MOF composites. [Reproduced with permission (Zhou et al. 2017). Copyright 2017, American Chemical Society].

Generally, among the carbon materials graphene oxide (GO) is widely adopted for humidity sensing due to its hydrophilic properties, cost-effectiveness, and low weight. In detail, QCMs exclusive of GO, containing a 166.5 µm thickness AT-cut quartz crystal inserted by two Au probes was used for the fabrication purpose (Jin et al. 2017). The GO diluted by deionized water dispersion was drop-casted on the sensing surface and dried. The sensor showed a response of ~1371/1% RH over 10–60% RH (as per Q factor) and 1068 Hz/10% RH at 70% RH as per frequency). The sensor was fast with rise and decay time of 20 and 3 s, correspondingly. It must be noted that the sensitivity depends

Table 7.8 Summary of quartz crystal microbalance-based humidity sensors

Materials used	Deposition method	QCM fundamental frequency (MHz)	Detection range (%RH)	Sensor sensitivity	Response/ recovery time	References
GO	Drop-casting	10	60–70	106.8 Hz/% RH	20 s/3 s	(Jin et al. 2017)
rGO/Polyethylene Oxide (PEO)	Spray method	10	11.3–84	20 Hz/% RH	11 s/7 s	(Wang et al. 2018)
GO/PDDAC	LBL self-assembly	8	0–97	25.4 Hz/% RH	10 s/10 s	(Ren et al. 2018)
GO/PANI/SnO$_2$	In-situ oxidative polymerization	8	0–97	29.1 Hz/% RH	7 s/2 s	(Zhang et al. 2018)
GO/Cu(OH)$_2$ NWs	GO drop-casting Cu NWs by sputtering	8	20–80	52.1 Hz/% RH	2 s/8 s	(Fang et al. 2020)
Graphene QD/Cs	Drop-casting	10	11–95	39.2 Hz/% RH	36/3 s	(Qi et al. 2019)
CNT-HKUST-1 composite	Spin coating technique	–	5–75	2.5×10^{-5}/% RH	250 s/265 s	(Chappanda et al. 2018)
MIL-101-derived hollow ball-like TiO$_2$	Spray coating	8	0–97	33.8 Hz/% RH	5 s/2 s	(Zhang et al. 2019)
GO/ethyleneimine	Spray method	–	11.3–97.3	27.3 Hz/1% RH	53 s/18 s	(Yuan et al. 2016)
Anodized alumina	Electron beam deposition	5	27–59	2 Hz/% RH	5 s/5 s	(Yamamoto et al. 2019)
BiOCl	Drop-casting	10	11–97	7.3 Hz/% RH	5.2 s/4.5 s	(Chen et al. 2020)
Black phosperous nanosheets	Drop-casting	10	11.3–84.3	Change in resonance frequency: 863 Hz (QCM-2 μL BP)	14 s/10 s	(Yao et al. 2017)
Cellulose nanocrystals	Spraying process	10	11.3–97.3	275/(54.3–97.3 % RH)	60 s/15 s	(Yao et al. 2020)
Nitro-modified cellulose nanocrystals	Drop casting	–	11–84	25.6 Hz/% RH	18 s/10 s	(Tang et al. 2021)
Copper (II)-MOF	–	–	17.2–97.6	28.7 Hz/% RH	30 s/18 s	(Zhou et al. 2017)

upon the thickness of graphene oxide layers. The superhydrophilic $Cu(OH)_2$ nanowires/graphene oxide nanocomposites were used for the humidity sensing. The $Cu(OH)_2$ nanowires acted as a hydrophilic material and therefore, easily absorbed with other hydrophilic materials, like GO (Fang et al. 2020). The combined geometry greatly improved the sensing performance in terms of response, fast detection, and reversibility. The material was coated using a drop-casting method. Different samples were prepared and it was observed that samples such as QCM-G, QCM-0, QCM-1, QCM-5, QCM-10, QCM-15, and QCM-20 showed corresponding frequency shifts around 601 Hz, 995 Hz, 1620 Hz, 3580 Hz, 4170 Hz, 7700 Hz and 6965 Hz to 80% RH exposure. Sensor response time and recovery times were greatly improved. The sensor's respiratory performance was also analyzed, and therefore, this could be useful in practical applications.

Another class of materials like metal organic compounds (MOFs) was widely investigated for humidity sensing due to its highly porous structure, which is a periodic network structure composed of inorganic metal centers. In this, the MOF along with metal oxides offer improvement in the sensitivity. In detail, the MOF composed of 3 types of TiO_2 structures like a hollow ball, nanosphere, and nanoflower based QCM humidity sensor was reported (Zhang et al. 2019). The corresponding frequency shift to 97% RH for hollow ball, nanoflower, and nanosphere TiO_2 QCM sensors around 1493, 1838, and 3286 Hz was observed. Among these samples, a ball-like TiO_2 layered device displayed a good response to the humidity of (0–97% RH), due to its large surface area and thus, created more active sites for effective adsorption of water. This study put a new insight into the decoration of new materials into MOF for humidity sensing. A new type of material like bismuth oxychloride (BiOCl) was also tested for humidity sensing (Chen et al. 2020). This material has good optical, catalytic, and electrical properties, and therefore, is considered for humidity sensing. This material was synthesized by hydrolysis method and drop cast onto the QCM chip. The change in frequency after 80% RH did not follow a linear trend. Also, the sensitivity of 7.3 Hz/% RH was limited. Among the materials, cellulose has a high specific surface area, distributed structure, hygroscopic dielectric material, and good heat resistant properties, and therefore, adopted for humidity sensing. In short, a nitro-modified cellulose nanocrystals (NCNCs) based material was used for humidity sensing owing to its polar hydroxyl and nitro groups which provided more adsorption sites for water molecules (Tang et al. 2021). For different samples of QCM-1, QCM-2, QCM-3, QCM-4, and QCM-5 the frequency shifts of −599 Hz, −844 Hz, −1728 Hz, −1868 Hz, and −2038 Hz at 84% RH, were observed. Here, the substitution of the nitro group in cellulose is important.

7.9 CONCLUSION AND OUTLOOK

Humidity detection is very important for various applications like environmental, biological, and medical, etc. This chapter reviews the most recent, advanced, and reliable techniques for the humidity sensing. These techniques include the key operating principle, design parameters, materials structures, sensing mechanism, advantages and limitations. It was observed that no single method can offer a complete sensing performance to the current problems and different applicability of humidity sensing in various applications. However, most of the methods discussed in this chapter offered great advantages over the conventional methods. The techniques such as an electrochemical, resistive, capacitive, optical fiber, and quartz crystal microbalance are widely adopted for humidity sensing.

The highlights of this chapter are mentioned below:

1. A lot of electrochemical humidity sensors based on metal oxides, 2D materials, and polymers have been reported. The 2D materials show high detection bandwidth. The metal-substituted MOXs also display an effective sensing performance. However, suitable addition needs to be done for better performance. In the electrochemical sensor, the porous materials have great applicability due to their high surface area. In this, an ordered porous network and large pore volume could effectively evade the agglomeration.

2. The capacitive sensors are based on the change in the dielectric property under the adsorption of humidity on the sensing layer. The sensor exhibited a wide detection range and good repeatability. It is very simple, easy to fabricate, inexpensive, and has less power consumption. But these sensors show that the materials performance gets affected by temperature and interfering species. Mostly polymers are extensively used in sensor fabrication. The material's like MOFs have a great future in this area. Also, this method is not suitable in a strong electromagnetic noisy or corrosive environments.

3. The resistive type sensors are dependent upon the change in the material resistance due to the interaction of humidity adsorption. These are a low-cost and sensitive types and offer good selectivity. However, this sensor exhibits certain limitations like the impact of temperature.

4. Fiber optics principle is mostly dependent upon the change in the refractive index of the materials, and therefore more reliable. It is a widely used approach for humidity due to its fast

response, ability to integrate with any type of molecule, real-time and multiplexing ability. Certain parameters like controlling the thickness of coated material, the fiber mechanical strength, and cross-interfering effects from different gases need to be considered.

5. The mass-sensitive method like QCM exhibits a principle of frequency shift due to humidity change. It is very simple, low weight and easy setup for humidity sensing, and therefore, extensively used for industrial applications. For this type of technique, the material with a low weight and hydrophilic property is highly preferred.

6. An integrated type of sensing approach like resistive/capacitive is highly required to fit different purposes of applications and environmental conditions.

7. The hydrophilic polymer cellulose materials have great choice to improve the performance of the device. The nanograined materials have higher adsorption capability than the micro grained materials, and therefore, research should focus in that direction as well. The porous materials could help in improving the sensing response due to their high surface area and thus, can be used in the nanocomposite form.

8. The development of self-powered materials for humidity sensing could be a futuristic task. Such devices can also be coupled to wearable devices and thus, can be used in smart sensing devices for health monitoring applications.

9. Future progress will be the development of a very miniaturized, low-weight, flexible, multifunctional, and reliable sensor for humidity monitoring.

REFERENCES

Aguiar, Maurício F. de, Andressa N.R. Leal, Celso P. de Melo and Kleber G.B. Alves. 2021. Polypyrrole-coated electrospun polystyrene films as humidity sensors. Talanta 234: 122636. doi:https://doi.org/10.1016/j.talanta.2021.122636.

Aksu, Zeriş, Cengiz Han Şahin and Murat Alanyalıoğlu. 2022. Fabrication of janus GO/rGO humidity actuator by one-step electrochemical reduction route. Sensors and Actuators B: Chemical 354: 131198. doi:https://doi.org/10.1016/j.snb.2021.131198.

Al-Hayali, Sarah Kadhim, Ansam M. Salman and Abdul Hadi Al-Janabi. 2021. High sensitivity balloon-like interferometric optical fiber humidity sensor based on tuning gold nanoparticles coating thickness. Measurement 170: 108703. doi:https://doi.org/10.1016/j.measurement.2020.108703.

Alrammouz, R., J. Podlecki, A. Vena, R. Garcia, P. Abboud, R. Habchi and B. Sorli. 2019. Highly porous and flexible capacitive humidity sensor based on self-assembled graphene oxide sheets on a paper substrate. Sensors and Actuators B: Chemical 298: 126892. doi:https://doi.org/10.1016/j.snb.2019.126892.

Arrizabalaga, Oskar, Javier Velasco, Joseba Zubia, Idurre Sáez de Ocáriz and Joel Villatoro. 2019. Miniature interferometric humidity sensor based on an off-center polymer cap onto optical fiber facet. Sensors and Actuators B: Chemical 297: 126700. doi:https://doi.org/10.1016/j.snb.2019.126700.

Bian, Ce, Jie Wang, Xiaohong Bai, Manli Hu and Tingting Gang. 2020. Optical fiber based on humidity sensor with improved sensitivity for monitoring applications. Optics and Laser Technology 130: 106342. doi:https://doi.org/10.1016/j.optlastec.2020.106342.

Blank, T.A., L.P. Eksperiandova and K.N. Belikov. 2016. Recent trends of ceramic humidity sensors development: a review. Sensors and Actuators B: Chemical 228: 416–42. doi:https://doi.org/10.1016/j.snb.2016.01.015.

Blessi, S., A. Manikandan, S. Anand, M. Maria Lumina Sonia, V. Maria Vinosel, Abeer Mohamed Alosaimi, Anish Khan, et al. 2021. Enhanced electrochemical performance and humidity sensing properties of Al^{3+} substituted mesoporous SnO_2 nanoparticles. Physica E: Low-Dimensional Systems and Nanostructures 133: 114820. doi:https://doi.org/10.1016/j.physe.2021.114820.

Chappanda, K.N., Osama Shekhah, Omar Yassine, Shashikant P. Patole, Mohamed Eddaoudi and Khaled N. Salama. 2018. The quest for highly sensitive qcm humidity sensors: the coating of CNT/MOF composite sensing films as case study. Sensors and Actuators B: Chemical 257: 609–619. doi:https://doi.org/10.1016/j.snb.2017.10.189.

Chen, Qiao, Ning-bo Feng, Xian-he Huang, Yao Yao, Ying-rong Jin, Wei Pan and Dong Liu. 2020. Humidity-sensing properties of a biocl-coated quartz crystal microbalance. ACS Omega 5(30): 18818–18825. American Chemical Society. doi:10.1021/acsomega.0c01946.

Chen, Zhao-Chi, Tien-Li Chang, Kai-Wen Su, Hsin-Sheng Lee and Jung-Chang Wang. 2021a. Application of self-heating graphene reinforced polyvinyl alcohol nanowires to high-sensitivity humidity detection. Sensors and Actuators B: Chemical 327: 128934. doi:https://doi.org/10.1016/j.snb.2020.128934.

Chen, Mao-qing, Yong Zhao, He-ming Wei, Cheng-liang Zhu and Sridhar Krishnaswamy. 2021b. 3D printed castle style fabry-perot microcavity on optical fiber tip as a highly sensitive humidity sensor. Sensors and Actuators B: Chemical 328: 128981. doi:https://doi.org/10.1016/j.snb.2020.128981.

Chi, Hong, Lim Jun Ze, Xuemin Zhou and Fuke Wang. 2021. GO film on flexible substrate: an approach to wearable colorimetric humidity sensor. Dyes and Pigments 185: 108916. doi:https://doi.org/10.1016/j.dyepig.2020.108916.

Chung, C.K., C.A. Ku and Z.E. Wu. 2021. A high-and-rapid-response capacitive humidity sensor of nanoporous anodic alumina by one-step anodizing commercial 1050 aluminum alloy and its enhancement mechanism. Sensors and Actuators B: Chemical 343: 130156. doi:https://doi.org/10.1016/j.snb.2021.130156.

Das, Sagnik, Md Lutfor Rahman, Partha P. Mondal, Preeti L. Mahapatra and Debdulal Saha. 2021. Screen-printed $MgAl_2O_4$ semi-thick film based highly sensitive and stable capacitive humidity sensor. Ceramics International 47 (23): 33515–33524. doi:https://doi.org/10.1016/j.ceramint.2021.08.260.

Dwiputra, Muhammad Adam, Farah Fadhila, Cuk Imawan and Vivi Fauzia. 2020. The enhanced performance of capacitive-type humidity sensors based on ZnO nanorods/WS_2 nanosheets heterostructure. Sensors and Actuators B: Chemical 310: 127810. doi:https://doi.org/10.1016/j.snb.2020.127810.

Fang, Han, Jianbin Lin, Zhixiang Hu, Huan Liu, Zirong Tang, Tielin Shi and Guanglan Liao. 2020. $Cu(OH)_2$ nanowires/graphene oxide composites based QCM humidity sensor with fast-response for real-time respiration monitoring. Sensors and Actuators B: Chemical 304: 127313. doi:https://doi.org/10.1016/j.snb.2019.127313.

Fard, Z.H., Y. Kalinovskyy, D.M. Spasyuk, B.A. Blight and G.K.H. Shimizu. 2016. Alkaline-earth phosphonate mofs with reversible hydration-dependent fluorescence. Chemical Communications 52(87): 12865–12868. The Royal Society of Chemistry. doi:10.1039/C6CC06490F.

Feng, Yu, Shijing Gong, Erwei Du, Ke Yu, Jie Ren, Zhenguo Wang and Ziqiang Zhu. 2019. TaS_2 nanosheet-based ultrafast response and flexible humidity sensor for multifunctional applications. Journal of Materials Chemistry C 7(30): 9284–9292. The Royal Society of Chemistry. doi:10.1039/C9TC02785H.

Fernandez, Frances Danielle M., Murali Bissannagari and Jihoon Kim. 2021. Fully inkjet-printed $BaTiO_3$ capacitive humidity sensor: microstructural engineering of the humidity sensing layer using bimodal ink. Ceramics International 47(17): 24693–24698. doi:https://doi.org/10.1016/j.ceramint.2021.05.191.

Gao, Yuan, Pengtao Jing, Ning Yan, Michiel Hilbers, Hong Zhang, Gadi Rothenberg and Stefania Tanase. 2017. dual-mode humidity detection using a lanthanide-based metal–organic framework: towards multifunctional humidity sensors. Chemical Communications 53(32): 4465–4468. The Royal Society of Chemistry. doi:10.1039/C7CC01122A.

Guo, Huayang, Changyong Lan, Zhifei Zhou, Peihua Sun, Dapeng Wei and Chun Li. 2017. Transparent, flexible, and stretchable WS_2 based humidity sensors for electronic skin. Nanoscale 9(19): 6246–6253. The Royal Society of Chemistry. doi:10.1039/C7NR01016H.

Hammouche, H., H. Achour, S. Makhlouf, A. Chaouchi and M. Laghrouche. 2021. A comparative study of capacitive humidity sensor based on keratin film, keratin/graphene oxide, and keratin/carbon fibers. Sensors and Actuators A: Physical 329: 112805. doi:https://doi.org/10.1016/j.sna.2021.112805.

He, Jiang, Peng Xiao, Jiangwei Shi, Yun Liang, Wei Lu, Yousi Chen, Wenqin Wang, et al. 2018. High performance humidity fluctuation sensor for wearable devices via a bioinspired atomic-precise tunable graphene-polymer heterogeneous sensing junction. Chem. Mater. 30: 4343–4354. doi:10.1021/acs.chemmater.8b01587.

He, Chenyang, Serhiy Korposh, Ricardo Correia, Liangliang Liu, B.R. Hayes-Gill and Stephen P. Morgan. 2021. Optical fibre sensor for simultaneous temperature and relative humidity measurement: towards absolute humidity evaluation. Sensors and Actuators B: Chemical 344: 130154. doi:https://doi.org/10.1016/j.snb.2021.130154.

He, Jing, Xiaotong Zheng, Zhiwen Zheng, Degang Kong, Kai Ding, Ningjun Chen, Haitao Zhang, et al. 2022. Pair directed silver nano-lines by single-particle assembly in nanofibers for non-contact humidity sensors. Nano Energy 92: 106748. doi:https://doi.org/10.1016/j.nanoen.2021.106748.

Hromadka, Jiri, Nurul N. Mohd Hazlan, Francisco U. Hernandez, Ricardo Correia, A. Norris, Stephen P. Morgan and Sergiy Korposh. 2019. Simultaneous in situ temperature and relative humidity monitoring in mechanical ventilators using an array of functionalised optical fibre long period grating sensors. Sensors and Actuators B: Chemical 286: 306–314. doi:https://doi.org/10.1016/j.snb.2019.01.124.

Huo, Yanming, Miaomiao Bu, Zongtao Ma, Jingyao Sun, Yuhua Yan, Kunhao Xiu, Ziying Wang, et al. 2022. Flexible, non-contact and multifunctional humidity sensors based on two-dimensional phytic acid doped co-metal organic frameworks nanosheets. Journal of Colloid and Interface Science 607: 2010–2018. doi:https://doi.org/10.1016/j.jcis.2021.09.189.

Ji, Cuiping, Jing Zeng, Sijia Qin, Min Chen and Limin Wu. 2021a. Angle-independent responsive organogel retroreflective structural color film for colorimetric sensing of humidity and organic vapors. Chinese Chemical Letters 32(11): 3584–3590. doi:https://doi.org/10.1016/j.cclet.2021.03.058.

Ji, Peng, Meng Zhu, Changrui Liao, Cong Zhao, Kaiming Yang, Cong Xiong, Jinli Han, et al., 2021b. In-fiber polymer microdisk resonator and its sensing applications of temperature and humidity. ACS Applied Materials & Interfaces 13(40): 48119–48126. American Chemical Society. doi:10.1021/acsami.1c14499.

Jin, Hao, Xiang Tao, Bin Feng, Liyang Yu, Demiao Wang, Shurong Dong and Jikui Luo. 2017. A humidity sensor based on quartz crystal microbalance using graphene oxide as a sensitive layer. Vacuum 140: 101–105. doi:https://doi.org/10.1016/j.vacuum.2016.10.017.

Kalyakin, A.S., A.N. Volkov and M.Yu. Gorshkov. 2020. An electrochemical sensor based on zirconia and calcium zirconate electrolytes for the inert gas humidity analysis. Journal of the Taiwan Institute of Chemical Engineers 111: 222–227. doi:https://doi.org/10.1016/j.jtice.2020.02.009.

Kalyakin, Anatoly S., Nikolai A. Danilov and Alexander N. Volkov. 2021. Determining humidity of nitrogen and air atmospheres by means of a protonic ceramic sensor. Journal of Electroanalytical Chemistry 895: 115523. doi:https://doi.org/10.1016/j.jelechem.2021.115523.

Kanaparthi, Srinivasulu. 2017. Pencil-drawn paper-based non-invasive and wearable capacitive respiration sensor. Electroanalysis 29(12): 2680–2684. John Wiley & Sons, Ltd. doi:https://doi.org/10.1002/elan.201700438.

Kano, Shinya, Kwangsoo Kim and Minoru Fujii. 2017. Fast-response and flexible nanocrystal-based humidity sensor for monitoring human respiration and water evaporation on skin. doi:10.1021/acssensors.7b00199.

Kano, Shinya and Minoru Fujii. 2018. All-painting process to produce respiration sensor using humidity-sensitive nanoparticle film and graphite trace. ACS

Sustainable Chemistry & Engineering 6(9): 12217–12223. American Chemical Society. doi:10.1021/acssuschemeng.8b02550.

Khalifa, Mohammed, Guenter Wuzella, Herfried Lammer and A.R. Mahendran. 2020. Smart paper from graphene coated cellulose for high-performance humidity and piezoresistive force sensor. Synthetic Metals 266: 116420. doi:https://doi.org/10.1016/j.synthmet.2020.116420.

Kim, Seulki, Sung G. Han, Young G. Koh, Hyunjung Lee and Wonmok Lee. 2018. Colorimetric humidity sensor using inverse opal photonic gel in hydrophilic ionic liquid. Sensors. doi:10.3390/s18051357.

Kim, Il Jin, Ji Eun Lee, Jae Wang Ko, Jiyoon Jung, Albert S. Lee, Dong Jin Lee, Seung Geol Lee, et al. 2020. Molecularly engineered copolyimide film for capacitive humidity sensor. Materials Letters 268: 127565. doi:https://doi.org/10.1016/j.matlet.2020.127565.

Kou, Donghui, Wei Ma, Shufen Zhang, Jodie L. Lutkenhaus and Bingtao Tang. 2018. High-performance and multifunctional colorimetric humidity sensors based on mesoporous photonic crystals and nanogels. Research-article. ACS Applied Materials & Interfaces 10. American Chemical Society: 41645–41654. doi:10.1021/acsami.8b14223.

Li, Jinze, Jianqi Zhang, Hao Sun, Yixin Yang, Yunlong Ye, Juntao Cui, Weiming He, et al. 2020. An optical fiber sensor based on carboxymethyl cellulose/carbon nanotubes composite film for simultaneous measurement of relative humidity and temperature. Optics Communications 467: 125740. doi:https://doi.org/10.1016/j.optcom.2020.125740.

Liu, LiangLiang, Stephen P. Morgan, Ricardo Correia and Serhiy Korposh. 2022. A single-film fiber optical sensor for simultaneous measurement of carbon dioxide and relative humidity. Optics & Laser Technology 147: 107696. doi:https://doi.org/10.1016/j.optlastec.2021.107696.

Lopez-Torres, Diego, Cesar Elosua, Joel Villatoro, Joseba Zubia, Manfred Rothhardt, Kay Schuster and Francisco J. Arregui. 2017. Enhancing sensitivity of photonic crystal fiber interferometric humidity sensor by the thickness of SnO_2 thin films. Sensors and Actuators B: Chemical 251: 1059–1067. doi:https://doi.org/10.1016/j.snb.2017.05.125.

Mahapatra, Preeti Lata, Sagnik Das, Partha Pratim Mondal, Tanushri Das, Debdulal Saha and Mrinal Pal. 2021. Microporous copper chromite thick film based novel and ultrasensitive capacitive humidity sensor. Journal of Alloys and Compounds 859: 157778. doi:https://doi.org/10.1016/j.jallcom.2020.157778.

Mallick, Shoaib, Zubair Ahmad, Karwan Wasman Qadir, Abdul Rehman, R.A. Shakoor, Farid Touati and S.A. Al-Muhtaseb. 2020. Effect of $BaTiO_3$ on the sensing properties of PVDF composite-based capacitive humidity sensors. Ceramics International 46(3): 2949–2953. doi:https://doi.org/10.1016/j.ceramint.2019.09.291.

Meng, Yahui, Yunfeng Cao, Hairui Ji, Jie Chen, Zhibin He, Zhu Long and Cuihua Dong. 2020. Fabrication of environmental humidity-responsive iridescent films with cellulose nanocrystal/polyols. Carbohydrate Polymers 240: 116281. doi:https://doi.org/10.1016/j.carbpol.2020.116281.

Mergu, Naveen, Hyorim Kim, Jiwon Ryu and Young-A Son. 2020. A simple and fast responsive colorimetric moisture sensor based on symmetrical conjugated polymer. Sensors and Actuators B: Chemical 311: 127906. doi:https://doi.org/10.1016/j.snb.2020.127906.

Niu, Hongsen, Wenjing Yue, Yang Li, Feifei Yin, Song Gao, Chunwei Zhang, Hao Kan, et al. 2021. Ultrafast-response/recovery capacitive humidity sensor based on arc-shaped hollow structure with nanocone arrays for human physiological signals monitoring. Sensors and Actuators B: Chemical 334: 129637. doi:https://doi.org/10.1016/j.snb.2021.129637.

Norton, Amie E., Mahmood Karimi Abdolmaleki, Daoli Zhao, Stephen D. Taylor, Steven R. Kennedy, Trevor D. Ball, Mark O. Bovee, et al. 2022. Vapoluminescence hysteresis in a Platinum(II) salt-based humidity sensor: mapping the vapochromic response to water vapor. Sensors and Actuators B: Chemical 359: 131502. doi:https://doi.org/10.1016/j.snb.2022.131502.

Pang, Yu, Jinming Jian, Tao Tu, Zhen Yang, Jiang Ling, Yuxing Li, Xuefeng Wang, et al. 2018. Wearable humidity sensor based on porous graphene network for respiration monitoring. Biosensors and Bioelectronics 116: 123–129. doi:https://doi.org/10.1016/j.bios.2018.05.038.

Parangusan, Hemalatha, Jolly Bhadra, Zubair Ahmad, S. Mallick, Farid Touati and Noora Al-Thani. 2020. Capacitive type humidity sensor based on PANI decorated Cu–ZnS porous microspheres. Talanta 219: 121361. doi:https://doi.org/10.1016/j.talanta.2020.121361.

Qi, Pengjia, Tong Zhang, Jieren Shao, Bai Yang, Teng Fei and Rui Wang. 2019. A QCM humidity sensor constructed by graphene quantum dots and chitosan composites. Sensors and Actuators A: Physical 287: 93–101. doi:https://doi.org/10.1016/j.sna.2019.01.009.

Qi, Rongrong, Tong Zhang, Xin Guan, Jianxun Dai, Sen Liu, Hongran Zhao and Teng Fei. 2020. Capacitive humidity sensors based on mesoporous silica and poly(3,4-Ethylenedioxythiophene) composites. Journal of Colloid and Interface Science 565: 592–600. doi:https://doi.org/10.1016/j.jcis.2020.01.062.

Qiang, Tian, Cong Wang, Ming-Qing Liu, Kishor Kumar Adhikari, Jun-Ge Liang, Lei Wang, Yang Li, et al. 2018. High-performance porous MIM-type capacitive humidity sensor realized via inductive coupled plasma and reactive-ion etching. Sensors and Actuators B: Chemical 258: 704–714. doi:https://doi.org/10.1016/j.snb.2017.11.060.

Qin, Lan, Sha Zhou, Yan Zhou and Lei Han. 2019. A naphthalenediimide-based Co-MOF as naked-eye colorimetric sensor to humidity. Journal of Solid State Chemistry 277: 658–664. doi:https://doi.org/10.1016/j.jssc.2019.07.030.

Ren, X., D. Zhang, D. Wang, Z. Li and S. Liu. 2018. Quartz crystal microbalance sensor for humidity sensing based on layer-by-layer self-assembled PDDAC/graphene oxide film. IEEE Sensors Journal 18(23): 9471–9476. doi:10.1109/JSEN.2018.2872854.

Sahoo, Karunakar, Biswajyoti Mohanty, Amrita Biswas and Jhasaketan Nayak. 2020. Role of hexamethylenetetramine in ZnO-Cellulose nanocomposite enabled UV and humidity sensor. Materials Science in Semiconductor Processing 105: 104699. doi:https://doi.org/10.1016/j.mssp.2019.104699.

Shrivastav, Anand M., Dinusha S. Gunawardena, Zhengyong Liu and H.-Y. Tam. 2020. Microstructured optical fiber based fabry–pérot interferometer as a humidity sensor utilizing chitosan polymeric matrix for breath monitoring. Scientific Reports 10(1): 6002. doi:10.1038/s41598-020-62887-y.

Shukla, S.K., Chandra Shekhar Kushwaha, Ayushi Shukla and G.C. Dubey. 2018. Integrated approach for efficient humidity sensing over zinc oxide and polypyrole composite. Materials Science & Engineering C, No. 2017. Elsevier B.V. doi:10.1016/j.msec.2018.04.054.

Sikarwar, S. and B.C. Yadav. 2015. Opto-electronic humidity sensor: a review. Sensors and Actuators A: Physical 233: 54–70. doi:https://doi.org/10.1016/j.sna.2015.05.007.

Sobhanimatin, Mohammad Bagher, S. Pourmahdian and Md. Mehdi Tehranchi. 2021. Colorimetric monitoring of humidity by opal photonic hydrogel. Polymer Testing 98: 106999. doi:https://doi.org/10.1016/j.polymertesting.2020.106999.

Su, Pi-Guey and Xin-Han Lee. 2018. Electrical and humidity-sensing properties of flexible metal-organic framework M050(Mg) and KOH/M050 and AuNPs/ M050 composites films. Sensors and Actuators B: Chemical 269: 110–117. doi:https://doi.org/10.1016/j.snb.2018.05.002.

Su, P., W. Liu, Y. Hong, Y. Ye and S. Huang. 2022. Vapor deposition of ultrathin hydrophilic polymer coatings enabling candle soot composite for highly sensitive humidity sensors. Materials Today Chemistry 24: 100786. doi:https://doi.org/10.1016/j.mtchem.2022.100786.

Sun, Chengyuan, Dandan Zhu, Haiyan Jia, Kun Lei, Zhen Zheng and X. Wang. 2019. Humidity and heat dual response cellulose nanocrystals/Poly(N-Isopropylacrylamide) composite films with cyclic performance. ACS Applied Materials & Interfaces 11(42): 39192–39200. American Chemical Society. doi:10.1021/acsami.9b14201.

Sunilkumar, A., S. Manjunatha, B. Chethan, Y.T. Ravikiran and T. Machappa. 2019. Polypyrrole–tantalum disulfide composite: an efficient material for fabrication of room temperature operable humidity sensor. Sensors and Actuators A: Physical 298: 111593. doi:https://doi.org/10.1016/j.sna.2019.111593.

Švara Fabjan, Erika, Peter Nadrah, Anja Ajdovec, Matija Tomšič, Goran Dražić, Matjaž Mazaj, Nataša Zabukovec Logar, et al. 2020. Colorimetric cutoff indication of relative humidity based on selectively functionalized mesoporous silica. Sensors and Actuators B: Chemical 316: 128138. doi:https://doi.org/10.1016/j.snb.2020.128138.

Tachibana, Shogo, Yi-Fei Wang, Tomohito Sekine, Yasunori Takeda, Jinseo Hong, Ayako Yoshida, Mai Abe, et al. 2022. A printed flexible humidity sensor with high sensitivity and fast response using a cellulose nanofiber/carbon black composite. ACS Applied Materials & Interfaces 14(4): 5721–5728. American Chemical Society. doi:10.1021/acsami.1c20918.

Tang, Lirong, Weixiang Chen, Bo Chen, Rixin Lv, Xinyu Zheng, Cheng Rong, Beili Lu, et al. 2021. Sensitive and renewable quartz crystal microbalance humidity sensor based on nitrocellulose nanocrystals. Sensors and Actuators B: Chemical 327: 128944. doi:https://doi.org/10.1016/j.snb.2020.128944.

Tian, Entao, Jingxia Wang, Yongmei Zheng, Yanlin Song, Lei Jiang and Daoben Zhu. 2008. Colorful humidity sensitive photonic crystal hydrogel. Journal of Materials Chemistry 18(10): 1116–1122. The Royal Society of Chemistry. doi:10.1039/B717368G.

Tong, Ruijie, Yu Wang, Kai-jun Zhao, Xiang Li and Yong Zhao. 2022. Surface plasmon resonance optical fiber sensor for relative humidity detection without temperature crosstalk. Optics & Laser Technology 150: 107951. doi:https://doi.org/10.1016/j.optlastec.2022.107951.

Trachioti, Maria G. and Mamas I. Prodromidis. 2020. Humidity impedimetric sensor based on vanadium pentoxide xerogel modified screen–printed graphite electrochemical cell. Talanta 216: 121003. doi:https://doi.org/10.1016/j.talanta.2020.121003.

Tsai, Y.-T., C.-W. Wu, L. Tsai and C.-C. Chiang. 2021. Application of graphene oxide-based, long-period fiber grating for sensing relative humidity. Journal of Lightwave Technology 39(12): 4124–4130. doi:10.1109/JLT.2020.3006380.

Vivekananthan, V., N.R. Alluri, Y. Purusothaman, A. Chandrasekhar, S. Selvarajan and Sang-Jae Kim. 2018. Biocompatible collagen nanofibrils: an approach for sustainable energy harvesting and battery-free humidity sensor applications. ACS Applied Materials & Interfaces 10(22): 18650–18656. American Chemical Society. doi:10.1021/acsami.8b02915.

Wang, Si, Guangzhong Xie, Yuanjie Su, Lei Su, Qiuping Zhang, Hongfei Du, Huiling Tai, et al. 2018. Reduced graphene oxide-polyethylene oxide composite films for humidity sensing via quartz crystal microbalance. Sensors and Actuators B: Chemical 255: 2203–2210. doi:https://doi.org/10.1016/j.snb.2017.09.028.

Wang, Yunlong, Yunqi Liu, Fang Zou, Chen Jiang, Chengbo Mou and Tingyun Wang. 2019a. Humidity sensor based on a long-period fiber grating coated with polymer composite film. Sensors 19(10): 2263. https://doi.org/10.3390/s19102263.

Wang, Zhihao, Yihe Zhang, Wenjiang Wang, Qi An and Wangshu Tong. 2019b. High performance of colorimetric humidity sensors based on minerals. Chemical Physics Letters 727: 90–94. doi:https://doi.org/10.1016/j.cplett.2019.04.066.

Wang, Xu-dong and Otto S. Wolfbeis. 2020. Fiber-optic chemical sensors and biosensors (2015–2019). Analytical Chemistry 92(1): 397–430. American Chemical Society. doi:10.1021/acs.analchem.9b04708.

Wang, Yang, Lina Zhang, Jinping Zhou and Ang Lu. 2020. Flexible and transparent cellulose-based ionic film as a humidity sensor. ACS Applied Materials & Interfaces 12(6): 7631–7638. American Chemical Society. doi:10.1021/acsami.9b22754.

Wang, Jia-Kai, Yu Ying, Nan Hu and Si-Yu Cheng. 2021. Double D-shaped optical fiber temperature and humidity sensor based on ethanol and polyvinyl alcohol. Optik 242: 166972. doi:https://doi.org/10.1016/j.ijleo.2021.166972.

Wen, Hsin-Yi, Yi-Ching Liu and Chia-Chin Chiang. 2020. The use of doped conductive bionic muscle nanofibers in a tennis racket–shaped optical fiber humidity sensor. Sensors and Actuators B: Chemical 320: 128340. doi:https://doi.org/10.1016/j.snb.2020.128340.

Wu, Jing-Xing and Bing Yan. 2017. A dual-emission probe to detect moisture and water in organic solvents based on green-Tb^{3+} post-coordinated metal–organic frameworks with red carbon dots. Dalton Transactions 46(21): 7098–7105. The Royal Society of Chemistry. doi:10.1039/C7DT01352C.

Wu, Jin, Yan-ming Sun, Zixuan Wu, Xin Li, Nan Wang, Kai Tao and G.P. Wang. 2019. Carbon nanocoil-based fast-response and flexible humidity sensor for multifunctional applications. Research-article. ACS Applied Materials & Interfaces 11: 4242–4251. American Chemical Society. doi:10.1021/acsami. 8b18599.

Wu, Zhilin, Jie Yang, Xia Sun, Yingjie Wu, Ling Wang, Gang Meng, Delin Kuang, et al. 2021. An excellent impedance-type humidity sensor based on halide perovskite $CsPbBr_3$ nanoparticles for human respiration monitoring. Sensors and Actuators B: Chemical 337: 129772. doi:https://doi.org/10.1016/j. snb.2021.129772.

Xia, Diandong, Jingfang Li, Weizuo Li, Lijun Jiang and Guangming Li. 2021. Lanthanides-based multifunctional luminescent films for ratiometric humidity sensing, information storage, and colored coating. Journal of Luminescence 231: 117784. doi:https://doi.org/10.1016/j.jlumin.2020.117784.

Xiang, Xinxin, Hui Ouyang, Jizhou Li and Zhifeng Fu. 2021. Humidity-sensitive $CsPbBr_3$ perovskite based photoluminescent sensor for detecting water content in herbal medicines. Sensors and Actuators B: Chemical 346: 130547. doi:https://doi.org/10.1016/j.snb.2021.130547.

Xu, Wei, Feiming Li, Zhixiong Cai, Yiru Wang, Feng Luo and Xi Chen. 2016. An ultrasensitive and reversible fluorescence sensor of humidity using perovskite $CH_3NH_3PbBr_3$. Journal of Materials Chemistry C 4(41): 9651–9655. The Royal Society of Chemistry. doi:10.1039/C6TC01075J.

Yamamoto, Nobuo, Tomiharu Yamaguchi and Kazuhiro Hara. 2019. Development of QCM humidity sensors using anodized alumina film. Electronics and Communications in Japan 102(11): 39–46. John Wiley & Sons, Ltd. doi:https://doi.org/10.1002/ecj.12218.

Yang, Haowei, Lei Pan, Yingping Han, Lihua Ma, Yao Li, Hongbo Xu and Jiupeng Zhao. 2017. A visual water vapor photonic crystal sensor with PVA/SiO_2 opal structure. Applied Surface Science 423: 421–425. doi:https://doi. org/10.1016/j.apsusc.2017.06.140.

Yao, Yao, Hui Zhang, Jie Sun, Wenying Ma, Li Li, Wenzao Li and Jiang Du. 2017. Novel QCM humidity sensors using stacked black phosphorus nanosheets as sensing film. Sensors and Actuators B: Chemical 244: 259–264. doi:https://doi.org/10.1016/j.snb.2017.01.010.

Yao, Xuesong and Yue Cui. 2020. A PEDOT:PSS functionalized capacitive sensor for humidity. Measurement 160: 107782. doi:https://doi.org/10.1016/j. measurement.2020.107782.

Yao, Yao, Xian-he Huang, Bo-ya Zhang, Zhen Zhang, Dong Hou and Ze-kun Zhou. 2020. Facile fabrication of high sensitivity cellulose nanocrystals based QCM humidity sensors with asymmetric electrode structure. Sensors and Actuators B: Chemical 302: 127192. doi:https://doi.org/10.1016/j.snb.2019. 127192.

Yeo, T.L., T. Sun and K.T.V. Grattan. 2008. Fibre-optic sensor technologies for humidity and moisture measurement. Sensors and Actuators A: Physical 144(2): 280–295. doi:https://doi.org/10.1016/j.sna.2008.01.017.

Yu, Jingjing, Francis Tsow, Sabrina Jimena Mora, Vishal Varun Tipparaju and Xiaojun Xian. 2021a. Hydrogel-incorporated colorimetric sensors with high humidity tolerance for environmental gases sensing. Sensors and Actuators B: Chemical 345: 130404. doi:https://doi.org/10.1016/j.snb.2021.130404.

Yu, Changgui, Huaping Gong, Zhaoxu Zhang, Kai Ni and Chunliu Zhao. 2021b. Optical fiber humidity sensor based on the vernier effect of the fabry-perot interferometer coated with PVA. Optical Fiber Technology 67: 102744. doi:https://doi.org/10.1016/j.yofte.2021.102744.

Yu, He, Jun-Ge Liang, Cong Wang, Cheng-Cai Liu, Bing Bai, Fan-Yi Meng, Dan-Qing Zou, et al. 2021c. Target properties optimization on capacitive-type humidity sensor: ingredients hybrid and integrated passive devices fabrication. Sensors and Actuators B: Chemical 340: 129883. doi:https://doi.org/10.1016/j.snb.2021.129883.

Yuan, Zhen, Huiling Tai, Zongbiao Ye, Chunhua Liu, Guangzhong Xie, X. Du and Yadong Jiang. 2016. Novel highly sensitive QCM humidity sensor with low hysteresis based on graphene oxide (GO)/Poly(Ethyleneimine) layered film. Sensors and Actuators B: Chemical 234: 145–154. doi:https://doi.org/10.1016/j.snb.2016.04.070.

Zhang, Ying, Bo Fu, Kuixue Liu, Yupeng Zhang, Xu Li, Shanpeng Wen, Yu Chen, et al. 2014. Humidity sensing properties of $FeCl_3$-NH_2-MIL-125(Ti) composites. Sensors and Actuators B: Chemical 201: 281–285. doi:https://doi.org/10.1016/j.snb.2014.04.075.

Zhang, Jingjing, Liang Sun, Chuan Chen, Man Liu, Wei Dong, Wenbin Guo and Shengping Ruan. 2017. High performance humidity sensor based on metal organic framework MIL-101(Cr) nanoparticles. Journal of Alloys and Compounds 695: 520–25. doi:https://doi.org/10.1016/j.jallcom.2016.11.129.

Zhang, Dongzhi, Dongyue Wang, Xiaoqi Zong, Guokang Dong and Yong Zhang. 2018. High-performance QCM humidity sensor based on graphene Oxide/Tin Oxide/Polyaniline ternary nanocomposite prepared by in-situ oxidative polymerization method. Sensors and Actuators B: Chemical 262: 531–541. doi:https://doi.org/10.1016/j.snb.2018.02.012.

Zhang, D., H. Chen, P. Li, D. Wang and Z. Yang. 2019. Humidity sensing properties of metal organic framework-derived hollow ball-like TiO_2 coated QCM sensor. IEEE Sensors Journal 19(8): 2909–2915. doi:10.1109/JSEN.2018.2890738.

Zhang, Jianxin, Xueyun Shen, Miao Qian, Zhong Xiang and Xudong Hu. 2021. An optical fiber sensor based on polyimide coated fiber bragg grating for measurement of relative humidity. Optical Fiber Technology 61: 102406. doi:https://doi.org/10.1016/j.yofte.2020.102406.

Zhang, Shiqi, Li Li, Yang Lu, Dapeng Liu, Junyao Zhang, Dandan Hao, Xuan Zhang, et al. 2022a. Sensitive humidity sensors based on ionically conductive metal-organic frameworks for breath monitoring and non-contact sensing. Applied Materials Today 26: 101391. doi:https://doi.org/10.1016/j.apmt.2022.101391.

Zhang, Xinran, Jiekai Lv, Jingshi Liu, Shihan Xu, Jiao Sun, Lin Wang, Lin Xu, et al. 2022b. Stable EMT type Zeolite/CsPbBr$_3$ perovskite quantum dot nanocomposites for highly sensitive humidity sensors. Journal of Colloid and Interface Science 616: 921–928. doi:https://doi.org/10.1016/j.jcis.2022.02.079.

Zhang, Zhaoxu, Huaping Gong, Changgui Yu, Kai Ni and Chunliu Zhao. 2022c. An optical fiber humidity sensor based on femtosecond laser micromachining fabry-perot cavity with composite film. Optics & Laser Technology 150: 107949. doi:https://doi.org/10.1016/j.optlastec.2022.107949.

Zhao, Hongran, Tong Zhang, Rongrong Qi, Jianxun Dai, Sen Liu and Teng Fei. 2017a. Drawn on paper: a reproducible humidity sensitive device by handwriting. ACS Applied Materials & Interfaces 9(33): 28002–28009. American Chemical Society. doi:10.1021/acsami.7b05181.

Zhao, Jing, Na Li, Hua Yu, Zheng Wei, Mengzhou Liao, Peng Chen, Shuopei Wang, et al. 2017b. Highly sensitive MoS$_2$ humidity sensors array for noncontact sensation. Advanced Materials 29(34): 1702076. John Wiley & Sons, Ltd. doi:https://doi.org/10.1002/adma.201702076.

Zhao, Yalei, Bin Yang and Jingquan Liu. 2018. Effect of interdigital electrode gap on the performance of SnO$_2$-modified MoS$_2$ capacitive humidity sensor. Sensors and Actuators B: Chemical 271: 256–263. doi:https://doi.org/10.1016/j.snb.2018.05.084.

Zhao, Qiuni, Zhen Yuan, Zaihua Duan, Yadong Jiang, Xian Li, Zhemin Li and Huiling Tai. 2019. An ingenious strategy for improving humidity sensing properties of multi-walled carbon nanotubes via Poly-L-Lysine modification. Sensors and Actuators B: Chemical 289: 182–185. doi:https://doi.org/10.1016/j.snb.2019.03.070.

Zhou, Zhuoqiang, Ming-Xing Li, Luyu Wang, Xiang He, Tao Chi and Zhao-Xi Wang. 2017. Antiferromagnetic Copper(II) metal–organic framework based quartz crystal microbalance sensor for humidity. Crystal Growth & Design 17(12): 6719–6124. American Chemical Society. doi:10.1021/acs.cgd.7b01318.

Zhou, Cheng, Qian Zhou, Bo Wang, Jiajun Tian and Yong Yao. 2021. High-sensitivity relative humidity fiber-optic sensor based on an internal-external Fabry-Perot cavity Vernier effect. Optics Express 29(8): 11854–11868. OSA. doi:10.1364/OE.421060.

Zhu, Penghui, Huajie Ou, Yudi Kuang, Lijing Hao, Jingjing Diao and Gang Chen. 2020. Cellulose nanofiber/carbon nanotube dual network-enabled humidity sensor with high sensitivity and durability. ACS Applied Materials & Interfaces 12(29): 33229–33238. American Chemical Society. doi:10.1021/acsami.0c07995.

pH Sensor

8.1   INTRODUCTION

The detection of hydrogen ion concentration, i.e. pH is always crucial for various applications including biomedical, marine science, civil engineering, life science, and industrial applications (Steinegger et al. 2020). A traditional glass electrode-based pH sensor is extensively used due to its simplicity, wide measuring range, high long-term stability, and mechanically constancy. However, it suffered from the problems like highly brittle, large impedance, and shrinking problems (Mu et al. 2021). Nowadays, various other methods such as electrochemical, field effect transistors, and optical fiber have been widely used in the monitoring of pH with great accuracy.

Several materials such as metal coated metal oxides (MOX), carbon, and polymers have been explored for pH sensing (Alam et al. 2018; Manjakkal et al. 2020). However, the limitation of MOX is a sluggish response in equally neutral and basic solutions. The limitation can be addressed by adopting a composite strategy (Gill et al. 2008; Manjakkal et al. 2020). The indicator dyes are also widely utilized for pH sensing due to their easy availability and stability. However, most of the dye's detection range is limited. This problem could be resolved by using a combined mixture of dyes. Table 8.1 shows numerous pH sensitive materials along with their advantages and disadvantages.

This chapter reviews different pH-sensitive techniques integrated with advanced materials structures along with their merits and demerits. The future outlook for the development of materials-based pH-sensing technology is outlined.

Table 8.1 Various materials along with their advantages and limitations for pH sensing

Materials	Advantages	Limitations
Metal oxide	1. Good and fast response, long-term stability, high-selectivity and large bandwidth 2. Low drift and biocompatible 3. Low-cost and small size	1. Slow response in both neutral and basic solutions 2. Response is dependent upon the size, shape, and morphology
Metal/metal oxides	1. Improve sensing performance 2. Chemically stable, less affected by salt solution	1. Limited detection range 2. Poor resolution and repeatability
Polymer	1. Flexible and stretchable 2. High conductivity and biocompatible	1. Limited detection range 2. Poor reliability
Carbon	1. Good response 2. Easily coated on flexible a substrate	1. Limited detection range 2. Need defect free material and less reusability

8.2   ELECTROCHEMICAL-BASED pH SENSORS

The electrochemical biosensors give a simple and cost-effective approach for the monitoring of biological parameters like pH via a simple direct electrochemical reaction. In this section, we discussed some electrochemical methods used for the detection of pH. A few electrochemical-based pH sensors are shown in Figure 8.1. Table 8.2 shows a summary of some electrochemical pH sensors.

For many years, the metal oxides have been used for the development of an electrochemical mechanism-based pH sensor. The unique properties like low-cost synthesis, good stability, different morphological and structural property, and the ability to tune the surface area, etc. are greatly utilized for effective pH sensing. In detail, a miniaturized sensing area of 1 mm^2 was reported for pH sensing (Santos et al. 2014). In order to obtain flexibility, the sensing layer was deposited on the polyimide substrate. In the structure, the Ag/AgCl acted as a reference electrode, a Pt wire as a counter electrode, and an Au thin film acted like a working electrode. The WO$_3$ nanoparticles varied the proton's concentration in solution and displayed a sensitivity of −56.7 ± 1.3 mV/pH. Its repeatability was also tested over 3 cycles. However, the sensor showed a little higher response time of 23 to 28 s. A highly sensitive and broad pH bandwidth was realized by using a polyaniline polymer (Zhu et al. 2022). This material offered many advantages like flexibility, good conductance, and a large surface-to-volume ratio. It showed a response of 63.72 mV/pH in 1–12 pH. The sensor was analyzed by measuring the

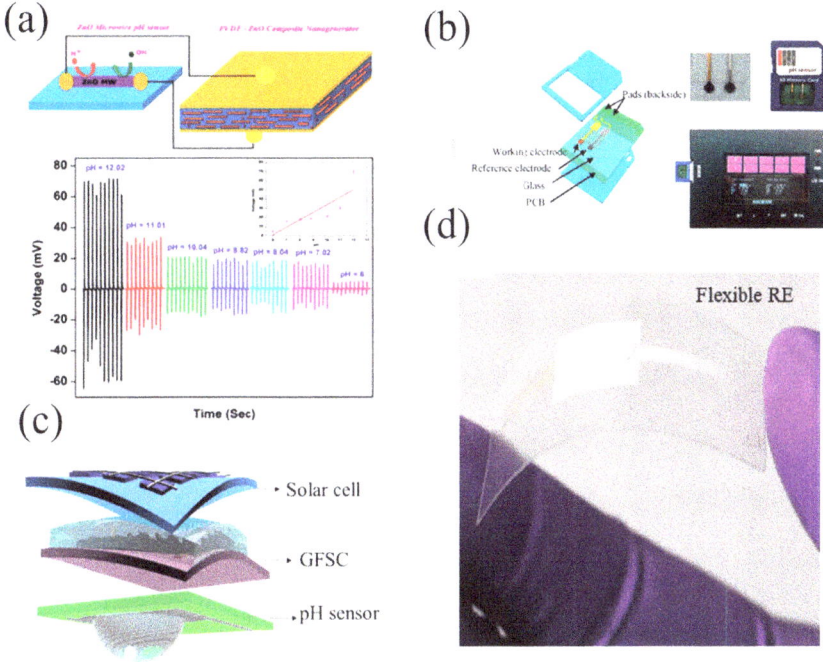

Figure 8.1 Numerous electrochemical-based pH sensing devices (a) pH sensor constructed by using ZnO microwires with the flexible organic-inorganic hybrid composite nanogenerator along its output, (b) An Ag/AgCl reference electrode coated with graphene oxide for pH sensing, (c) Graphene-Ag-3D graphene foam electrodes-based pH sensor, (d) Flexible Ag/AgCl/KCl used RE developed on PET substrate-based pH sensor. [Reproduced with permission (Manjakkal et al. 2020). Copyright 2020, Elsevier].

pH of real test samples like cola, orange, coffee, milk, tea, and soda and compared with commercial pH sensor. The results displayed that the RSD did not exceed 2.8%. The tryptophan (Trp) residues, along with Trp (Trp=O) and (Trp-OH) based potentiometer pH sensing devices were investigated (Hu et al. 2020). Generally, the proton combined electron transmission of Trp=O/Trp-OH was reliant on electrode base material. In the temperature study, it was observed that the rate of reaction between the Trp=O/Trp-OH decreased with increasing incubation temperature, this might be due to the weakening of binding. The sensor was tested to measure the pH of real samples like water, milk, and cola samples with much fewer interfering effects. The sensor sensitivity of 52 mV/pH was not too high and was observed closer to the theoretical sensitivity.

Graphene has attracted significant attention for the development of electrochemical-based pH sensing owing to its mesmerizing properties like high electron movement, thermal conduction, and ultra-high

Table 8.2 A summary of a few electrochemical-based pH sensors

Materials used	Detection range	Response time	Sensitivity (mV/pH)	References
WO_3 nanoparticle	5–9	23–28 s	−56.7	(Santos et al. 2014)
PANI	2–12	<10 s	58.2	(Yoon et al. 2017)
PANI	2–10	1.5 s	70.23	(Zhu et al. 2022)
Methylene blue	4.59–7.86	–	−56	(González-Fernández et al. 2022)
Schiff base polymer ($SBP_{PPDA\text{-}PDA}$)	2–9	–	57.3	(LI et al. 2021)
Nickel	9.4–12	–	2.3097 μA/pH	(Jafari et al. 2021)
Black phosphorus	1–8	–	$I_p = -0.9775$ pH + 8.4325	(Yi et al. 2020)
Trp (Trp=O) and its re-reduced species (Trp-OH)	1–12	–	52	(Hu et al. 2020)
PANI	4–7	–	−62.5	(Nyein et al. 2016)
Graphene-PANI (Gr-PANI)	1–11	–	−50.14 μA/pH·cm^2 in pH 1–5, and 139.2 μA/pH.cm^2 in pH 7–11	(Sha et al. 2017)
Poly (vinyl alcohol)/poly (acrylic acid) (PVA/PAA) hydrogel nanofibers	–	<2 s	74	(Shaibani et al. 2017)
SWCNTs	3–11	7 s	80 passes: −36.2 mV/pH	(Qin et al. 2016)
CuO nanorods	5–8.5		0.64 μF/pH at 50 Hz	(Manjakkal et al. 2018)
$RuO_2\text{-}Ta_2O_5$	2–12	<8 s and <15 s in acidic and basic solution	56	(Manjakkal et al. 2016)
MnO_2/GPLE	1.5–12.5	20 s in acidic medium and 60 s in alkaline medium	57.051	(Mohammad-Rezaei et al. 2019)
G-PU	6–9	5 s	47	(Manjakkal et al. 2019)

adsorption area. However, it is less selective to H^+ and thus, there is a decrease in EDL capacitive value. To improve its electrochemical performance, the graphene-PANI composite was used to detect the pH (Sha et al. 2017). In this work, use of such kind of sensing materials not only improved the detection range of 1–11 but also the response ~139.2 $\mu A/pHcm^2$ during pH 7–11. The drift rate and testing of real test samples was not considered. The electrochemical pH sensing was implemented by using a pencil lead electrode (PLE) exfoliated towards graphene nanolayers via using a potential (Mohammad-Rezaei et al. 2019). The PLE is cheap, electrically conductive and mechanically stable electrode. This structure was highly porous, has a large surface area and is conductive. The performance was greatly improved by addition of metal oxides such as manganese oxide nanoparticles. The experimental results of MnO_2/GPLE pH sensor confirmed the response of 57.051 mV/pH in 1.5–12.5. Over a period of 2 months, the device exhibited drift of just 7.6 mV. The sensor showed a low response time in acidic medium than in alkaline medium.

8.3 FIELD EFFECT TRANSISTOR-BASED pH SENSORS

Generally, an ion sensitive field effect transistor (ISFET) and extended gate field effect transistor (EGFET) have been widely adopted to pH monitoring because of their fast response, high resolution, sensitivity, and ease in preparation. ISFET device consistsof a source, gate, and drain terminals, in which gate material decides the sensitivity of the sensor. On other hand, the EGFET was modified with the reference electrode and a MOX semiconductor field effect transistor (MOSFET). The RE ensure stable signal across the MOSFET. Here, both types of FETs are discussed along with their limitations for pH sensing. A summary of a few field-effect transistor-based pH sensor is displayed in Table 8.3.

Table 8.3 A summary of a few field-effect transistor-based pH sensor

Type of confi-guration	Materials used	Substrate used	Detection range (%RH)	Sensitivity (mV/pH)	References
ISFET	ITO/polyethylene terephthalate (PET)	Polyethylene naphthalate (PEN)	4–10	−44.86	(Li et al. 2017)
EGFET	CuS	Glass	2–12	27.8	(Sabah et al. 2017)
ISFET	Silicon nanowire	Silicon-on-insulator (SOI)	3–10	1438.8	(Cho and Cho 2021)
FET	Silicon nanowire	Silicon	4–10	720.7	(Zhou et al. 2020)

(Contd.)

190

Table 8.3 A summary of a few field-effect transistor-based pH sensor (*Contd.*)

Type of configuration	Materials used	Substrate used	Detection range (%RH)	Sensitivity (mV/pH)	References
EGFET	TaO$_x$/Ta	Silicon	2–12	61.28	(Pan et al. 2022)
EGFET	PdO	Si/SiO$_2$/Pt	2–12	42.36	(Sharma et al. 2021)
MOSFET	Si/SiO$_2$	Silicon-on-insulator (SOI)	4–9	52.8	(Chen et al. 2022)
EGFET	TiO$_2$ Array-nanorods	Glass	2–12	78.25	(Khizir and Abbas 2022)
EGFET	TiO$_2$ Flower shaped- nanords	Glass	2–12	60.96	(Khizir and Abbas 2022)
EGFET	TiO$_2$ nanoflowers	FTO/glass	2–12	46	(Yang et al. 2019)
EGFET	Ga$_2$O$_3$ nanorod	ITO/glass	1–11	64.29	(Chiang et al. 2022)
EGFET	NiOx	Silicon	2–12	60.65	(Pan et al. 2021a)
EGFET	CuO Nanowire	–	2–12	48.34	(Mishra et al. 2020)
EGFET	ZnInxOy	PEN	2–12	61.76	(Pan et al. 2020)
ISFET	TiN	Alumina	4–10	53.98	(Sinha et al. 2021)
OECTs	PANI/PEDOT:PSS	Polyimide	4–10	100	(Demuru et al. 2021)
EGFET	CoNxOy	Silicon	4–10	64.36	(Wang and Pan 2021)
EGFET	Na$_3$BiO$_4$–Bi$_2$O$_3$	Indium–Tin-Oxide (ITO) coated glass	7–12	49.63	(Sharma et al. 2020)
ISFET	HfO$_2$/SiO$_2$	–	2–10	476.54	(Khwairakpam and Pukhrambam 2021)
EIS	LaTixOy	Silicon	2–12	68.17	(Pan et al. 2021b)
ISFET	Al$_2$O$_3$	–	4–9	347.6	(Kumar et al. 2020)
EGFET	Sn-doped 3D (micro/nano) V$_2$O$_5$ spheres	PSi	2–12	102	(Slewa et al. 2020)
EGFET	SiO$_2$	ITO-Glass	3–10	36.0	(Kang and Cho 2019)
	HfO$_2$		3–10	54.6	
	ZrO$_2$		3–10	55.5	
	Ta$_2$O$_5$		3–10	56.0	
	SnO$_2$		3–10	54.4	

The titanium/Au/Ag/silver chloride layered construction on a plastic substrate to pH sensing was reported (Li et al. 2017). This device showed very less voltage drift of <1.7 mV per hour. However, the complete details of the drift value were not provided. Recently, an ISFETs-based silicon nanowire synthesized by the electrospinning method was employed for pH sensing (Cho and Cho 2021). The processing procedure to fabricate PVP NFs on an SOI substrate is displayed in Figure 8.2(a). The sensor performance was studied by two methods of single-gate (SG) and dual-gate (DG) operation as shown in Figure 8.2(b). The sensor exhibited an improvement in sensitivity with the high capacitive-coupling ratios. The properties of silicon nanowire like high surface area, and higher gate capacitance benefitted to improve the sensing performance significantly and thus, showed a slope value of 1438.8 mV/pH. The sensor tested in the single-gate and dual-gate modes by using a film-type and SiNW-type and observed that the single-gate sensing mode performance was lower than the dual-gate mode.

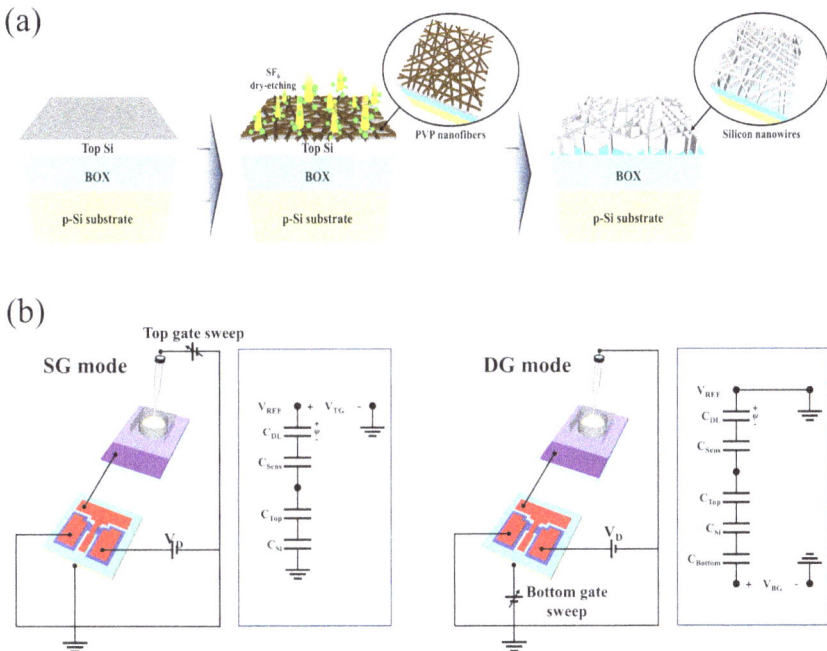

Figure 8.2 (a) Diagram showing the development of PVP NF fabrication on SOI substrate. (b) Different operation methods in SG mode and DG mode sensing. [Reproduced with permission (Cho and Cho 2021). Copyright 2020, Elsevier].

Based on dual intensification modes like capacitance and differentiation, the sensitivity of the pH sensor was greatly improved.

The silicon nanowire (SiNW) was used as a gate and the maximum potential of the device was intensified via the capacitance proportion of both the upper and down gates (Zhou et al. 2020). The differential amplification was done via a readout circuit chip. In this report, it was verified that the differential amplification might be tested to dual-gate devices as well. This sensor showed a response ~720.7 mV/pH to 4–10. Although, this pH range is limited and could limit its use for practical applications. In the cross-interfering study, the sensor response was deviated from the standard pH meter. The metal oxides structure was also used in FETs configuration for pH sensing (Sharma et al. 2021). By the e-beam evaporation method, the PdO was deposited onto the Si/SiO$_2$/Pt substrate. The PdO material was selected for sensing purposes due to its biocompatibility, high stability in both acidic and basic mediums, and offered good pH sensitivity. The sensor testing under real samples was not tested. Also, the sensor showed good linearity over a broad range but exhibited limited sensitivity. The sensitivity and pH detection range were improved by using hydrothermally synthesized Titanium-dioxide (TiO$_2$) nanorods MOXs nanomaterial for sensing (Khizir and Abbas 2022). This provided good stability and chemical stability, and well-played for pH detection. This device exhibited a sensitivity of 78.25 mV/pH, repeatability of 0.23%, and lesser hysteresis value of 9.1 mV, respectively. It was observed that the buffer solution played a crucial role in deciding the sensitivity. The interaction between the counter-ions to the buffered level increased the H ions activity. This sensor stability study was limited. The performance of sensing material needs to be checked in real test samples. Recently, gallium oxide (Ga$_2$O$_3$) nanorods were synthesized by CBD and used as a sensing membrane for pH sensing (Chiang et al. 2022). The Ga$_2$O$_3$ was chemically and thermally stable, nontoxic, and biocompatible. A variety of Ga$_2$O$_3$ nanorods were synthesized and thus, had different surface areas which benefitted to improve the performance. In this work, the pH detection range increased from 1–11, with a sensitivity of 64.29 mV/pH. The study also reported that at pH 4, 7, and 10 at RT for 6 h, the corresponding drift in potential was 3.47, 2.75, and 5.42 mV/h. However, the drift rate increased with the rise in pH value which may be due to an increase in OH$^-$ ions. A high-k materials like Ta$_2$O$_5$ can be effectively used to detect pH (Kang and Cho 2019). The other materials like SnO$_2$, HfO$_2$, ZrO$_2$ were also tested and it was observed that Ta$_2$O$_5$ displayed the highest response ~478.0 mV/pH with a drift rate ~24.6 mV/h. The Ta$_2$O$_5$ membrane displayed a minimum potential of 100.2 mV thru a pH loop of pH 4–pH 10. These sensing materials contained hydroxyl groups and thus, were very efficient for the effective adsorption of H ions. The response and corresponding linear regions toward different sensitive films in the top gate and bottom gate were also studied.

8.4 OPTICAL FIBER pH SENSORS

Optical fiber is very promising for the detection of various biological parameters due to its distinguished properties. In pH sensing, an optical fiber has proved to be very effective over other methods, as the effective refractive index of the coated material will get altered in presence of varied pH. Other advantages are a low-cost system, simple and easy fabrication, wide applications range, and tunability, etc. Here, some optical fiber pH sensors are discussed with their merits and demerits. Table 8.4 shows some development in fiber-optics based pH sensors.

Numerous polymers have been used for the detection of pH due to their sensitive nature, ease in preparation, cost-effectiveness, and refractive index compatibility with the optical fiber. For example, PANI a conducting polymer was elected for the pH sensing because it changes absorption properties (Khanikar and Singh 2019). The PANI was dispersed into the PVA, acted as a known steric stabilizer and provided stability. The combined structure was deposited onto the decladed part of the MMF fiber by drop-casting method and the transmission power under varied pH was measured. The response was measured from pH 2–9 range, and obtained a response of 2.79 μW/pH. The sensor reproducibility, stability, durability, long-term stability and effect of temperature was studied. Despite this simple and ease in sensor fabrication, the precaution in terms of bending, stress on fiber, etc., must be taken to avoid the impact on its absorbance. The fiber-optic sensors showed a response in terms of wavelength shifting because of modification in the effective refractive index of the material under varied pH and thus, more reliable than the absorbance-based sensors. The sensitivity and response time were significantly improved by changing the measuring method. In detail, the rGO-PANI structure was deposited on unclad part of the PCS-MMF, and measured its SPR dip wavelength under varied pH (Semwal and Gupta 2019). The PANI conductivity was varied under variation of pH. The rGO was very sensitive to both acidic and basic mediums and also its conductivity can be easily tuned under varied pH. Therefore, the sensor exhibited a good response of 75.09 nm/pH over 2.4 to 11.35 pH. However, the sensor showed an exponential nature. The sensor repeatability was checked just over 2 cycles. The fiber grating structure was also effectively used for the determination of pH change. In short, a long-period fiber grating (LPFG) coated with smart hydrogel and measured its performance under a pH range of 2–12 (Mishra et al. 2017). Under pH variation, the volume of the hydrogel changed and thus, the refractive index, which resulted in the change in the dip position of LPG spectrum. The response ~0.66 nm/pH with a response time of <2 s was obtained, and therefore, the sensor was effective for pH sensing with broad bandwidth. Furthermore, the effect of temperature

Table 8.4 A Summary of a few optical fiber-based pH sensors

Type of optical fiber	Materials used	Deposition method	Sensor sensitivity	Detection range	Response/ Recovery time (s)	References
Etched MMF	PANI-PVA	Dip coating	2.79 µW/pH	2–9	~8 s/22 s	(Khanikar and Singh 2019)
Etched multimode fibers (MMF)	SiO$_2$ and thick Au-SiO$_2$	Dip coating	19.9 T%/pH, 10.02 T%/pH	8–13	–	(Lu et al. 2021)
Uncladded PCS fiber	SiO$_2$-TiO$_2$	–	0.14 a.u. – 1.98 a.u.	1–12	0.31 s	(Alshoaibi and Islam 2021)
Uncladded plastic clad silica (PCS) multimode optical fiber	rGO-PANI	Dip coating	75.09 nm/pH	2.4–11.35	–	(Semwal and Gupta 2019)
Plastic multi-mode optical fiber	8-hydroxypyrene-1,3,6-trisulfonic acid trisodium salt (HPTS)	–	6–42.1 mV/pH	2.5–9	10 s	(Moradi et al. 2019)
Uncladded plastic clad silica (PCS) optical fiber	Phenol red encapsulated anatase nanoparticles	Dip coating	1.83 counts/pH	1–12	1.09 s/ 0.42 s	(Islam 2021)
Decladed PCS (Plastic Clad Silica) optical fiber	Polyethanol glycol assisted gold nanodendrites (AuNDs)	Drop-casting	24.95 count/pH	1–12	~0.87 s	(Islam et al. 2019)
LPG (SMF)	PAAm	Dip coating	0.66 nm/pH	2–12	2 s	(Mishra et al. 2017)
Silica fiber coated	Meta-cresol purple in an optimized TEOS/DDS	Dip coating	S = 35	7.4–9.7	2.5 min 6.5 min	(Chen et al. 2021)
MMF end coated	Hydrogel	Polymerization	–	5.5–8	30 s/-	(Gong et al. 2020)
Optical fiber coating	Naphth-AlkyneOMe	–	–	11–13	100 s	(Tariq et al. 2021)
Plastic Clad Silica decladed coated	Cresol red (CR) immobilized anatase nanoparticles (CR-ANPs)	–	7.05 counts/pH	1–7	–	(Islam and Alshoaibi 2021)
MMF decladed	Au-SiO$_2$	Dip coating	13.4 T%/pH	8–13	10 s	(Lu et al. 2021)
MZI tapered	DNA (i-motif)	–	480 pm/pH	4.98–7.4	–	(Wang et al. 2021)
FBG-MMF coated on cladding	PAA/CS	Layer by layer self-assembly technique	-9.29 nm/pH	4.71–8.52	40 min	(Li et al. 2022)

change on sensor performance was also investigated. However, the LPGs are delicate and, therefore, need to be handled properly.

Various pH-indicating dyes are also been widely used for pH sensing. However, it needs to incorporate into the certain host materials. In short, a meta-cresol purple (mCP) was immobilized into the TEOS/DDS sol-gel matrix (Chen et al. 2021). This sensor principle was dependent upon the interaction of EW with the entrapped mCP in the sensing layer as shown in Figure 8.3(a). This material was coated by a dip coating onto the optical fiber surface and the end tip coated by a silver mirror. The reflected light was again entrapped inside the region and increased the sensor response. The sensor performance is measured from acidic-to-basic and from basic-to-acidic conditions as shown in Figure 8.3(b). As this sensor was designed for monitoring the pH of the marine environments of pH in the range of 7.4–9.7 in seawater. But the sensor response time was a little longer in minutes. Furthermore, the sensor lifetime was 7 days.

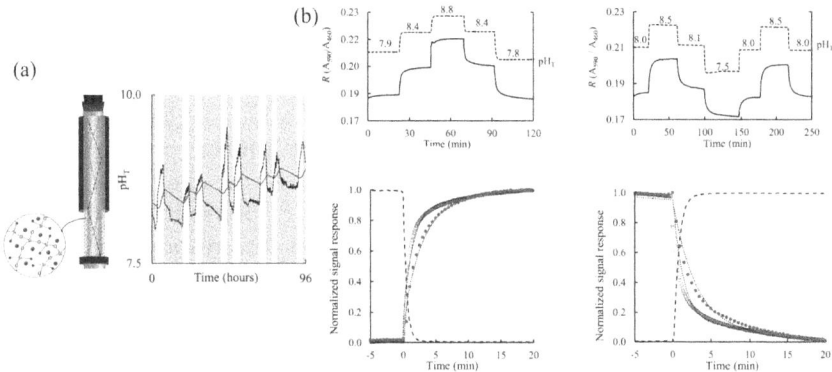

Figure 8.3 (a) Optical fiber pH sensor based on meta-cresol purple (mCP) was immobilized into the TEOS/DDS sol-gel matrix. (b) pH sensor response toward pH shift in ASW with sol-gel matrix film thickness. [Reproduced with permission (Chen et al. 2021). Copyright 2020, Elsevier].

The tapered fiber-based on the Mach-Zehnder interferometer was used for the pH detection. The middle section of the fiber was tapered by a flame scanning technique and it was functionalized by a complementary-DNA (Wang et al. 2021). It observed the device showed a linearity to 5–5 to 7.4 pH with a resolution of 0.042 in pH unit. Still, the response of the sensor was limited. However, this study proposed a new way in using of a DNA like material for pH sensing. Recently, a new type of approach for pH sensing along with the temperature compensation was provided. For example, a PAA/CS sensible coating developed on the surface of a fiber and the FBGs was used only for the temperature compensation (Li et al. 2022). To achieve the SPR effect, a gold layer was deposited onto the region of FBGs. It was observed that the dip corresponding to the surface coated with PAA/CS film significantly shifted, while there was no change in the FBGs peak wavelength. This suggested that refractive indices of the PAA/CS layer were pH related and showed a response of −9.29 nm/pH. This type of sensor has great potential in biosensing applications.

8.5   CONCLUSION AND OUTLOOK

Significant advances have been made in designing and demonstrating new pH sensing technologies for various applications. The successful integration of nanoscale materials such as MOXs, organic polymers, hydrogels, and carbon-based materials for detecting a very wide pH range on a smart sensing platform like electrochemical, field effect transistors, and optical fiber with enhanced performance is briefly summarized.

In an electrochemical technique, a significant change in the electrochemical potential occurs due to an ion interchange among sensible coatings and electrolytes. Ideally, the response of the sensor should be nearer to the Nernstian value (59.14 mV/pH at 25 °C). The most important part of the electrochemical sensor was the glass membrane electrode mostly made from MOXs, metallic/MOXs, polymeric and carbon-based. The MOXs-based pH sensors are considered because of their high sensitivity, ease in preparation, fast response in seconds, and small drift. In the case of MOXs-based sensors, the response is mainly dependent upon the kind of material, deposition method, structural, morphological, homogeneousness, and crystallinity, etc. Various fabrication methods like screen print, sputter, sol-gel, and CVD have been used. The thick layers have more advantages as compared to thin films glassy relied pH devices in the form of development of multiple detection arrays on a singly used substrate and cheap. Particular focus needs to be paid to the fabrication methods and material structure to improve the sensor performance.

Field effect-based transistors for pH sensing has attracted considerable attention due to their small size, fast and accurate detection, flexible, and label-free detection. The diffusion time is dependent upon the parameters such as flow rate and the size/shape of the fluidic channel. The hysteresis depends upon the thickness and multilayer of the film. The material's properties like size and structure also affect the sensing performance. Therefore, fewer defects and a controlled uniform morphological sensing membrane are ideal for efficient pH sensing.

In the fiber optics approach, the sensing layer consists of a pH-sensitive layer and fiber, and the change in PH will induce a change in the optical properties of the fiber. This approach is relatively simple and sensitive. It has many advantages over others due to its resistant property to harsh chemicals, and being highly safe in an explosive environments. The coating thickness of the sensing layer, and the proportion of indicator in host matrix need to be controlled for better sensing performance.

The other key parameters for the development of pH sensors are addressed below:

1. It is very important to focus on the material's toxicity, flexibility, and getting SE and RE on the same substrate.

2. A lot of metal oxides exhibit sluggish performance in the neutral and basic solutions. It can be resolved by adopting a method of mixing with the conducting material. Certain metal oxides like ZnO, Ta_2O_5, or SnO_2 can be effectively used because of biocompatibility and high sensitivity, and flexibility.

3. It is also observed that the structure of material properties like grain dimensions, material morphological features, surface-to-volume ratio, structural forms, and geometry modulate the sensing performance.

4. The future includes the development of a biocompatible, wearable, and flexible pH sensor for health monitoring. The tracking of diseases through the pH sweat of the person and using a smart sensory platform could be possible. This type of sensor can be used for water and food quality applications.

REFERENCES

Alam, Arif Ul, Y. Qin, S. Nambiar, John T.W. Yeow, Matiar M.R. Howlader, Nan-Xing Hu and M. Jamal Deen. 2018. Polymers and organic materials-based pH sensors for healthcare applications. Progress in Materials Science 96: 174–216. doi:https://doi.org/10.1016/j.pmatsci.2018.03.008.

Alshoaibi, Adil and Shumaila Islam. 2021. Mesoporous nanostructures-based fiber optic pH sensors: synthesis, structure-tailoring, physiochemical and sensing stimuli. Materials Research Bulletin 140: 111332. doi:https://doi.org/10.1016/j.materresbull.2021.111332.

Chen, Wan-Har, Wayne D.N. Dillon, Evelyn A. Armstrong, Stephen C. Moratti and Christina M. McGraw. 2021. Self-referencing optical fiber pH sensor for marine microenvironments. *Talanta* 225: 121969. doi:https://doi.org/10.1016/j.talanta.2020.121969.

Chen, Shulin, Yan Dong, Tzu-Li Liu and Jinghua Li. 2022. Waterproof, flexible field-effect transistors with submicron monocrystalline Si nanomembrane derived encapsulation for continuous pH sensing. Biosensors and Bioelectronics 195: 113683. doi:https://doi.org/10.1016/j.bios.2021.113683.

Chiang, Jung-Lung, Yi-Guo Shang, Bharath Kumar Yadlapalli, Fei-Peng Yu and Dong-Sing Wuu. 2022. Ga_2O_3 nanorod-based extended-gate field-effect transistors for pH sensing. Materials Science and Engineering: B 276: 115542. doi:https://doi.org/10.1016/j.mseb.2021.115542.

Cho, Seong-Kun and Won-Ju Cho. 2021. Ultra-high sensitivity pH-sensors using silicon nanowire channel dual-gate field-effect transistors fabricated by electrospun polyvinylpyrrolidone nanofibers pattern template transfer. Sensors and Actuators B: Chemical 326: 128835. doi:https://doi.org/10.1016/j.snb.2020.128835.

Demuru, Silvia, Brince Paul Kunnel and Danick Briand. 2021. Thin film organic electrochemical transistors based on hybrid PANI/PEDOT:PSS active layers for enhanced pH sensing. Biosensors and Bioelectronics: X 7: 100065. doi:https://doi.org/10.1016/j.biosx.2021.100065.

Gill, Edric, Khalil Arshak, Arousian Arshak and Olga Korostynska. 2008. Mixed metal oxide films as pH sensing materials. Microsystem Technologies 14(4): 499–507. doi:10.1007/s00542-007-0435-9.

Gong, Jingjing, Michael G. Tanner, Seshasailam Venkateswaran, James M. Stone, Yichuan Zhang and Mark Bradley. 2020. A hydrogel-based optical fibre fluorescent pH sensor for observing lung tumor tissue acidity. Analytica Chimica Acta 1134: 136–143. doi:https://doi.org/10.1016/j.aca.2020.07.063.

González-Fernández, Eva, Matteo Staderini, Jamie R.K. Marland, Mark E. Gray, Ahmet Uçar, Camelia Dunare, Ewen O Blair, et al. 2022. In, vivo application of an implantable tri-anchored methylene blue-based electrochemical pH sensor. Biosensors and Bioelectronics 197: 113728. doi:https://doi.org/10.1016/j.bios.2021.113728.

Hu, Gengxin, Nanxi Li, Yuwei Zhang and Hong Li. 2020. A novel pH sensor with application to milk based on electrochemical oxidative quinone-functionalization of tryptophan residues. Journal of Electroanalytical Chemistry 859: 113871. doi:https://doi.org/10.1016/j.jelechem.2020.113871.

Islam, Shumaila, Hazri Bakhtiar, Madzlan Aziz, Saira Riaz, Md. Safwan Abd Aziz, Shahzad Naseem and Nada Elshikeri. 2019. Optically active phenolphthalein encapsulated gold nanodendrites for fiber optic pH sensing. Applied Surface Science 485: 323–331. doi:https://doi.org/10.1016/j.apsusc.2019.04.210.

Islam, Shumaila. 2021. Fast responsive anatase nanoparticles coated fiber optic pH sensor. Journal of Alloys and Compounds 850: 156246. doi:https://doi.org/10.1016/j.jallcom.2020.156246.

Islam, Shumaila and Adil Alshoaibi. 2021. Thermally and optically functionalized anatase nano-cavities based fiber optic pH sensor. Materials Research Bulletin 133: 111017. doi:https://doi.org/10.1016/j.materresbull.2020.111017.

Jafari, Behnaz, Madhivanan Muthuvel and Gerardine G. Botte. 2021. Nickel-based electrochemical sensor with a wide detection range for measuring hydroxyl ions and pH sensing. Journal of Electroanalytical Chemistry 895: 115547. doi:https://doi.org/10.1016/j.jelechem.2021.115547.

Kang, Joo-Won and Won-Ju Cho. 2019. Achieving enhanced ph sensitivity using capacitive coupling in extended gate FET sensors with various High-K sensing films. Solid-State Electronics 152: 29–32. doi:https://doi.org/10.1016/j.sse.2018.11.008.

Khanikar, Tulika and Vinod Kumar Singh. 2019. PANI-PVA composite film coated optical fiber probe as a stable and highly sensitive pH sensor. Optical Materials 88 (September 2018). Elsevier: 244–251. doi:10.1016/j.optmat. 2018.11.044.

Khizir, Hersh Ahmed and Tariq Abdul-Hameed Abbas. 2022. Hydrothermal synthesis of TiO_2 nanorods as sensing membrane for extended-gate field-effect transistor (EGFET) pH sensing applications. Sensors and Actuators A: Physical 333: 113231. doi:https://doi.org/10.1016/j.sna.2021.113231.

Khwairakpam, Dayananda Singh and Puspa Devi Pukhrambam. 2021. Sensitivity optimization of a double-gated ISFET pH-sensor with HfO_2/SiO_2 gate dielectric stack. Microelectronics Journal 118: 105282. doi:https://doi. org/10.1016/j.mejo.2021.105282.

Kumar, Narendra, Deepa Bhatt, Moitri Sutradhar and Siddhartha Panda. 2020. Interface mechanisms involved in A-IGZO based dual gate ISFET pH sensor using Al_2O_3 as the top gate dielectric. Materials Science in Semiconductor Processing 119: 105239. doi:https://doi.org/10.1016/j.mssp.2020.105239.

Li, Q., W. Tang, Y. Su, Y. Huang, S. Peng, B. Zhuo, S. Qiu, et al. 2017. Stable thin-film reference electrode on plastic substrate for all-solid-state ion-sensitive field-effect transistor sensing system. IEEE Electron Device Letters 38(10): 1469–1472. doi:10.1109/LED.2017.2732352.

Li, Yan-Yan, Yu-Xi Yang, Sha-Sha Hong, Yao Liu, Zhi Yang, Bin-Yu Zhao, Jian-Po Su, et al. 2021. An electrochemical sensor based on redox-active schiff base polymers for simultaneous sensing of glucose and pH. Chinese Journal of Analytical Chemistry 49(6): e21118–25. doi:https://doi.org/10.1016/ S1872-2040(21)60107-X.

Li, Xuegang, Pengqi Gong, Qiming Zhao, Xue Zhou, Yanan Zhang and Yong Zhao. 2022. Plug-in optical fiber spr biosensor for lung cancer gene detection with temperature and pH compensation. Sensors and Actuators B: Chemical 359: 131596. doi:https://doi.org/10.1016/j.snb.2022.131596.

Lu, Fei, Ruishu Wright, Ping Lu, Patricia C. Cvetic and Paul R. Ohodnicki. 2021. Distributed fiber optic pH sensors using sol-gel silica based sensitive materials. Sensors and Actuators B: Chemical 340: 129853. doi:https://doi. org/10.1016/j.snb.2021.129853.

Manjakkal, Libu, Krzysztof Zaraska, Katarina Cvejin, Jan Kulawik and Dorota Szwagierczak. 2016. Potentiometric RuO_2–Ta_2O_5 pH sensors fabricated using thick film and LTCC technologies. Talanta 147: 233–240. doi:https://doi.org/10.1016/j.talanta.2015.09.069.

Manjakkal, Libu, Bhuvaneshwari Sakthivel, Nammalvar Gopalakrishnan and Ravinder Dahiya. 2018. Printed flexible electrochemical pH sensors based on CuO nanorods. Sensors and Actuators B: Chemical 263: 50–58. doi:https:// doi.org/10.1016/j.snb.2018.02.092.

Manjakkal, Libu, Wenting Dang, Nivasan Yogeswaran and Ravinder Dahiya. 2019. Textile-based potentiometric electrochemical pH sensor for wearable applications. Biosensors. doi:10.3390/bios9010014.

Manjakkal, Libu, Dorota Szwagierczak and Ravinder Dahiya. 2020. Metal oxides based electrochemical pH sensors: current progress and future perspectives. Progress in Materials Science 109: 100635. doi:https://doi.org/10.1016/j. pmatsci.2019.100635.

Mishra, A.K., D.K. Jarwal, B. Mukherjee, A. Kumar, S. Ratan and S. Jit. 2020. CuO nanowire-based extended-gate field-effect-transistor (FET) for pH sensing and enzyme-free/receptor-free glucose sensing applications. IEEE Sensors Journal 20 (9): 5039–5047. doi:10.1109/JSEN.2020.2966585.

Mishra, S.K., B. Zou and K.S. Chiang. 2017. Wide-range pH sensor based on a smart- hydrogel-coated long-period fiber grating. IEEE Journal of Selected Topics in Quantum Electronics 23 (2): 284–288. doi:10.1109/JSTQE. 2016.2629662.

Mohammad-Rezaei, Rahim, Sahand Soroodian and Ghadir Esmaeili. 2019. Manganese oxide nanoparticles electrodeposited on graphenized pencil lead electrode as a sensitive miniaturized pH sensor. Journal of Materials Science: Materials in Electronics 30(3): 1998–2005. doi:10.1007/s10854-018-0471-5.

Moradi, Vahid, Mohsen Akbari and Peter Wild. 2019. A Fluorescence-based pH sensor with microfluidic mixing and fiber optic detection for wide range pH measurements. Sensors and Actuators A: Physical 297: 111507. doi:https://doi.org/10.1016/j.sna.2019.07.031.

Mu, Boyu, Guoqing Cao, Luwei Zhang, Yu Zou and Xinqing Xiao. 2021. Flexible wireless pH sensor system for fish monitoring. Sensing and Bio-Sensing Research 34: 100465. doi:https://doi.org/10.1016/j.sbsr.2021.100465.

Nyein, Hnin Yin Yin, Wei Gao, Ziba Shahpar, Sam Emaminejad, Samyuktha Challa, Kevin Chen, Hossain M. Fahad, et al. 2016. A wearable electro-chemical platform for noninvasive simultaneous monitoring of Ca^{2+} and pH. ACS Nano 10(7): 7216–7224. American Chemical Society. doi:10.1021/acsnano. 6b04005.

Pan, Tung-Ming, Yen-Hsiang Huang, Jim-Long Her, Bih-Show Lou and See-Tong Pang. 2020. Solution processed $ZnIn_xO_y$ sensing membranes on flexible PEN for extended-gate field-effect transistor pH sensors. Journal of Alloys and Compounds 822: 153630. doi:https://doi.org/10.1016/j.jallcom. 2019.153630.

Pan, Tung-Ming, C.-H. Lin and S.-T. Pang. 2021a. Structural and sensing characteristics of NiO_x sensing films for extended-gate field-effect transistor pH sensors. IEEE Sensors Journal 21(3): 2597–2603. doi:10.1109/JSEN.2020. 3027060.

Pan, Tung-Ming, Prabir Garu and Jim-Long Her. 2021b. Influence of Ti content on sensing performance of LaTixOy sensing membrane based electrolyte-insulator-semiconductor pH sensor. Materials Chemistry and Physics 269: 124774. doi:https://doi.org/10.1016/j.matchemphys.2021.124774.

Pan, Tung-Ming, Chen-Hung Lin and See-Tong Pang. 2022. Structural properties and sensing performance of TaOx/Ta stacked sensing films for extended-gate field-effect transistor pH sensors. Journal of Alloys and Compounds 903: 163955. doi:https://doi.org/10.1016/j.jallcom.2022.163955.

Qin, Yiheng, Hyuck-Jin Kwon, Ayyagari Subrahmanyam, Matiar M.R. Howlader, P. Ravi Selvaganapathy, Alex Adronov and M. Jamal Deen. 2016. Inkjet-printed bifunctional carbon nanotubes for pH sensing. Materials Letters 176: 68–70. doi:https://doi.org/10.1016/j.matlet.2016.04.048.

Sabah, Fayroz A., Naser M. Ahmed, Z. Hassan and Munirah Abdullah Almessiere. 2017. Influence of CuS membrane annealing time on the sensitivity of EGFET pH sensor. Materials Science in Semiconductor Processing 71: 217–225. doi:https://doi.org/10.1016/j.mssp.2017.07.001.

Santos, Lídia, Joana P. Neto, Ana Crespo, Daniela Nunes, Nuno Costa, Isabel M. Fonseca, Pedro Barquinha, et al. 2014. WO_3 nanoparticle-based conformable pH sensor. ACS Applied Materials & Interfaces 6(15): 12226–12234. American Chemical Society. doi:10.1021/am501724h.

Semwal, Vivek and Banshi D. Gupta. 2019. Highly sensitive surface plasmon resonance based fiber optic ph sensor utilizing rGO-pani nanocomposite prepared by in situ method. Sensors and Actuators B: Chemical 283: 632–642. doi:https://doi.org/10.1016/j.snb.2018.12.070.

Sha, R., K. Komori and S. Badhulika. 2017. Amperometric pH sensor based on graphene–polyaniline composite. IEEE Sensors Journal 17(16): 5038–5043. doi:10.1109/JSEN.2017.2720634.

Shaibani, Parmiss Mojir, Hashem Etayash, Selvaraj Naicker, Kamaljit Kaur and Thomas Thundat. 2017. Metabolic study of cancer cells using a pH sensitive hydrogel nanofiber light addressable potentiometric sensor. ACS Sensors 2(1): 151–156. American Chemical Society. doi:10.1021/acssensors.6b00632.

Sharma, Prashant, Sandeep Gupta, Rini Singh, Kanad Ray, S.L. Kothari, Soumendu Sinha, Rishi Sharma, et al. 2020. Hydrogen ion sensing characteristics of Na_3BiO_4–Bi_2O_3 mixed oxide nanostructures based EGFET pH sensor. International Journal of Hydrogen Energy 45(37): 18743–18751. doi:https://doi.org/10.1016/j.ijhydene.2019.07.252.

Sharma, Prashant, Rini Singh, Rishi Sharma, Ravindra Mukhiya, Kamlendra Awasthi and Manoj Kumar. 2021. Palladium-oxide extended gate field effect transistor as pH sensor. Materials Letters: X 12: 100102. doi:https://doi.org/10.1016/j.mlblux.2021.100102.

Sinha, Soumendu, Tapas Pal, D.a Kumar, R. Sharma, D. Kharbanda, P.K. Khanna and R. Mukhiya. 2021. Design, fabrication and characterization of TiN sensing film-based ISFET pH sensor. Materials Letters 304: 130556. doi:https://doi.org/10.1016/j.matlet.2021.130556.

Slewa, Lary H., Tariq A. Abbas and Naser M. Ahmed. 2020. Effect of Sn doping and annealing on the morphology, structural, optical, and electrical properties of 3D (Micro/Nano) V_2O_5 sphere for high sensitivity PH-EGFET sensor. Sensors and Actuators B: Chemical 305: 127515. doi:https://doi.org/10.1016/j.snb.2019.127515.

Steinegger, Andreas, Otto S. Wolfbeis and Sergey M. Borisov. 2020. Optical sensing and imaging of pH values: spectroscopies, materials, and applications. Chemical Reviews 120(22): 12357–12489. American chemical society. doi:10.1021/acs.chemrev.0c00451.

Tariq, Ayedah, Jalal Baydoun, C. Remy, Rasta Ghasemi, Jean Pierre Lefevre, Cédric Mongin, Alexandre Dauzères, et al. 2021. Fluorescent molecular probe based optical fiber sensor dedicated to pH measurement of concrete. Sensors and Actuators B: Chemical 327: 128906. doi:https://doi.org/10.1016/j.snb.2020.128906.

Wang, Chih-Wei and Tung-Ming Pan. 2021. Structural properties and sensing performances of CoNxOy ceramic films for EGFET pH sensors. Ceramics International 47(18): 25440–25448. doi:https://doi.org/10.1016/j.ceramint.2021.05.266.

Wang, Yujia, Hao Zhang, Yunxi Cui, Shaoxiang Duan, Wei Lin and Bo Liu. 2021. A complementary-DNA-enhanced fiber-optic sensor based on microfiber-assisted mach-zehnder interferometry for biocompatible pH sensing. Sensors and Actuators B: Chemical 332: 129516. doi:https://doi.org/10.1016/j.snb.2021.129516.

Yang, Chih-Chiang, Kuan-Yu Chen and Yan-Kuin Su. 2019. TiO$_2$ nano flowers based EGFET sensor for pH sensing. Coatings. doi:10.3390/coatings9040251.

Yi, Jinquan, Xiaoping Chen, Qinghua Weng, Ying Zhou, Zhizhong Han, Jinghua Chen and Chunyan Li. 2020. A simple electrochemical pH sensor based on black phosphorus nanosheets. Electrochemistry Communications 118: 106796. doi:https://doi.org/10.1016/j.elecom.2020.106796.

Yoon, Jo Hee, Kyung Hoon Kim, Nam Ho Bae, Gap Seop Sim, Yong-Jun Oh, Seok Jae Lee, Tae Jae Lee, et al. 2017. Fabrication of newspaper-based potentiometric platforms for flexible and disposable ion sensors. Journal of Colloid and Interface Science 508: 167–173. doi:https://doi.org/10.1016/j.jcis.2017.08.036.

Zhou, Kun, Zhida Zhao, Pengbo Yu and Zheyao Wang. 2020. Highly sensitive pH sensors based on double-gate silicon nanowire field-effect transistors with dual-mode amplification. Sensors and Actuators B: Chemical 320: 128403. doi:https://doi.org/10.1016/j.snb.2020.128403.

Zhu, Chonghui, Hua Xue, Hongran Zhao, Teng Fei, Sen Liu, Qidai Chen, Bingrong Gao, et al. 2022. A dual-functional polyaniline film-based flexible electrochemical sensor for the detection of pH and lactate in sweat of the human body. Talanta 242: 123289. doi:https://doi.org/10.1016/j.talanta.2022.123289.

Chapter **9**

Biosensor

9.1   INTRODUCTION

Biosensors have tremendous demand not only in the healthcare sector but also in food safety, agricultural, and environmental monitoring applications. A highly sensitive, selective and stable biosensor is always necessary for quality checking and the healthcare sector. A biosensor consists of a bioreceptor, transducer, analyte and acquisition unit to display. Every biosensor consists of a biological element. The bio-recognition elements include different biological entities like a cell, tissue, enzymes, aptamers, different types of antibodies, and nucleic acid, etc. Firstly, the bio identification part interacts with the test sample of interest and later generates a signal which is being tested (Wang and Wolfbeis 2020). Enzymes are proteins that plays the role of biocatalysts, and therefore, increase the rate of reaction. Aptamer are target specific and even small changes in analyte can deviate the binding, and therefore, widely used in biosensing. Proteins consisting of amino acids are used for detection purpose. Molecularly imprinted polymers (MIPs) are template shaped cavities in polymer matrix and highly stale against pressure, temperature and pH. Based on biosensing principle, different methods have been implemented such as optical, electrochemical, and mass-based. The optical technique is based on the totally internal reflection phenomenon, weakly guiding modes, and surface charge plasmonic. When the analyte interacts with the biorecognition element, this modulates the refractive indices and thus, changes the resonance dip in the spectrum which is easily monitored via a spectrometer (Kim et al. 2021). In an electrochemical approach, the electrical properties of the solution get changed due to chemical reactions. It consists of three basic electrode systems, i.e., a working electrode, a counter electrode, and a reference electrode, which are used to apply the current and keep the potential stable and thus, improve the selectivity and stability (Stefano et al. 2022). It is categorized based on conductance, potentiometer, and amperemeter.

This chapter reviews the rapid development and recent advances in the field of biosensing for the detection of various biochemical species. Firstly, the discussion is about the various methods such as electrochemical, field effect transistors, fiber optics, whispering gallery mode, colorimetry, and fluorescence-based biosensing platforms based on various biomolecules along with their functionalization methods for the detection of numerous biochemical species. Finally, the discussion is based on the biosensing characteristics, critical parameters, advantages, limitations, and future research scope.

9.2 ELECTROCHEMICAL BIOSENSORS

9.2.1 Amperometry-based Biosensors

The carbon-based materials such as graphene, rGO, and GO have been extensively used for the detection of biochemical compounds. These materials have many advantages such as larger specific surface area for biomolecule immobilization, a larger electrical conductivity, and also play a catalytic role in the biochemical reaction. The following few carbon-based materials are discussed for biosensing applications. The amperometry-based phenolic compounds catechol detection was proposed by Palanisamy et al (Palanisamy et al. 2017). The graphene was used due to its high conductivity. In this work, the graphene-cellulose microfibers (GR-CMF)-based carbon electrodes were used for catechol detection in the range of 0.2 to 209.7 µM. The high conductivity of graphene and biocompatibility of cellulose microfibers assisted in the formation of laccase on the composite, and therefore, enhanced the sensitivity and lowered the detection limit. Recently, carbon cloth and carbon paper were extensively used in biosensing applications. As they possessed unique properties like biocompatibility, rigidity, and flexibility which are very useful for sensing various analytes. Carbon paper is a fiber-like morphology that provides inflexibility, while carbon cloth is a woven shaped flexible structure. The copper-cobalt mixed oxides-carbon paper-based glucose sensor was fabricated by controlling the Cu/Co ratio deposited onto the electrodes by a magnetron sputtering (López-Fernández et al. 2020). The optimized thickness of Cu/Co ratio was 3:4 was very sensitive to glucose. It was observed that the catalytic positions of 4Co-(O)-Cu are functioning for the electrooxidation of glucose over the electrode surface. The sensitivity of metal oxides (MOXs) to biosensing can be improved by adding carbon nanostructures into it and fabrication of a heterostructure.

The combined geometry of MOXs and carbon nanostructure was also implemented for biosensing. The Co_3O_4/reduced graphene oxide/carbon cloth (Co_3O_4/rGO/CC) composite was successfully prepared and utilized for hydrogen peroxide detection (Zhang et al. 2020). The agglomeration of rGO was prevented by CC and fast electron transfer occurred between the material and analyte. Recently, metal organic frameworks (MOFs) are also been widely adopted in biosensing. The high surface area and abundance of active sites have attracted this material for biosensing applications. A portable, very simple configured, and sensitive nonenzymatic glucose sensor was fabricated by using a Co-MOF/CC/Paper material (Wei et al. 2021). The Co-MOF/CC interface provided many catalytic sites and a high surface area for glucose sensing in the concentration of 0.8 mM to 16 mM. The sensor performance was also compared with the commercial glucometer. The long-term stability was studied over 4 months. However, a decrease in stability to 60% after 120 days was observed. A few biosensors-based on amperometry method are shown in Table 9.1.

Table 9.1 Summary of a few amperometry-based biosensors

Materials used	Analyte	Detection range (μM)	LOD (μM)	References
GR-CMF/laccase/ SPCE	Catechol	0.20–209.70	85	(Palanisamy et al. 2017)
Carbon paper/ $Cu_xCo_yO_z$	Glucose	0–1000	0.105	(López-Fernández et al. 2020)
Carbon cloth/ Co_3O_4-rGO	Hydrogen peroxide	387–63,523	0.022	(Zhang et al. 2020)
CC/NiO	Glucose	5–2000	0.0075	
CC/Co(MOF)	Glucose	800–16,000	150	(Wei et al. 2021)
Glassy carbon electrode/COx/ Au/MWNT	Cholesterol	2 μM–1.4 mM	0.5	(Haritha et al. 2022)
Tyr-cell surface display system/ glassy carbon electrode	Bisphenol A	0.00001–0.1	0.00001	(Zhao et al. 2022)
Glassy carbon electrode/PtNP@ MXene-$Ti_3C_2T_x$	L-glutamate	10×10^3– 110×10^3 nM	0.45×10^3 nM	(Liu et al. 2021a)
EIS/$Ti_3C_2T_x$/ Aptamer	STX	1–200 nM	0.03 nM	(Ullah et al. 2021)

9.2.2 Other Electrochemical Techniques-based Biosensors

In this section, we will discuss other methods for the detection of numerous biochemical parameters. Few biosensors developed by using an electrochemical approach is shown in Figure 9.1. Table 9.2 tabulated the recent development of some electrochemical-based biosensors.

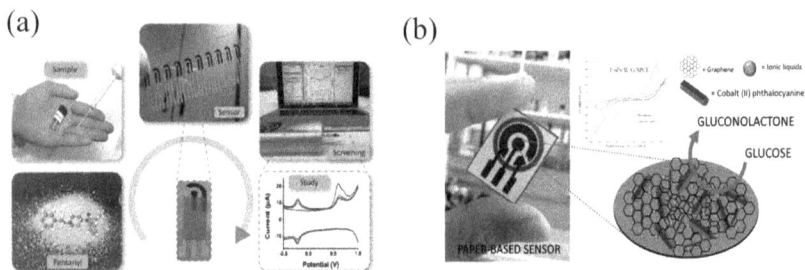

Figure 9.1 (a) Overview of fentanyl screening using SPCE-based electrochemical sensor analyzed by using a portable electro-analyzer. [Reproduced with permission (Goodchild et al. 2019). Copyright 2019, American Chemical Society]. (b) Paper-based glucose sensor based on CoPc/IL/G SPCE with PADs. [Reproduced with permission (Chaiyo et al. 2018). Copyright 2018, Elsevier].

A paper substrate-based 8-hydroxy-2'-deoxyguanosine (8-OHdG) electrochemical biosensor was reported by Martins et al. (Martins et al. 2017). The performance of the device was recorded by using the DPV technique. Conducting polymer PEDOT enhanced the oxidation peak current of 8-OHdG. The proposed sensor detected the 8-OHdG in 50–1000 ng/ml and the limit of detection (LOD) ~14.4 ng/m. The metal-coated metal oxides nanocomposite was effectively used for the detection of microRNA (miRNA). The Au nanomaterial conductivity and modified hairpin probe through Au-S bonds and the CeO_2 played a major role in electro-catalyze the H_2O_2 (Sun et al. 2018). The wide range of detection in the range of 1.0 fM–1000 fM with a LOD about 0.434 fM was achieved. The sensor of good linearity and selectivity to miRNA. The limitation of the current sensor was that the device could not measure the multi-targets simultaneously. The graphene-based network was developed using a laser-scribed graphene (LSG) to detect the glucose humanoid body liquids (whole blood, serum, sweat, and urine) (Prabhakaran and Nayak 2020). To increase catalytic efficacy, the MOXs CuO was used. The sensor performance was tested on curvilinear body parts. Nowadays, zeolites-based biosensors have attracted particular interest in sensor development due to their fascinating cation interchange and electrocatalytic features. The dopamine detection was performed by using Cu^{2+} rich "zeolite A"/nitrogen fixed graphene materials (Hatefi-Mehrjardi et al. 2019). The zeolite was prepared by a

Table 9.2 A summary of a few electrochemical-based biosensors

Materials used	Analyte	Measurement technique	Detection range	Limit of detection	References
PEDOT modified paper sensor	8-OHdG	DPV	50–1000 ng/mL	14.4 ng/mL	(Martins et al. 2017)
Paper based–(CeO$_2$–Au@GOx)	miR–21	DPV	1.0 fM–1000 fM	0.434 fM	(Sun et al. 2018)
Au/Ti$_3$C$_2$-AuNps	MicroRNA-155	DPV	1×10^{-6}–10 nM	0.35×10^{-6} nM	(Yang et al. 2020)
Onion Peel derived carbon nanospheres/CFP	Progesterone	DPV	37.39 pM–0.25 nM	0.012 nM	(Akshaya et al. 2019)
Chitosan/IL/GNS/Glassy carbon electrode	BSA	DPV	1.0×10^{-10}–1.0×10^{-4} g/L	2×10^{-11} g/L	(Xia et al. 2016)
Glassy carbon electrode/ hc-g-C$_3$N$_4$@CDs/h-FABP-Ab1/ BSA/h-FABP	Fatty acid–binding protein	DPV	0.66×10^{-6}–0.066×10^{-3} nM	0.22×10^{-6} nM	(Karaman et al. 2021)
MWCNTs-CS/Pd NPs/C$_{60}$/ MB-DNA/GCE.	Bisphenol A	DPV	1–50 μM	0.5 μM	(Jalalvand et al. 2019)
Glassy carbon electrode/ Ti$_3$C$_2$T$_x$/ChOx/Chitosan	Cholesterol	DPV	1–4.5 nM	0.11 nM	(Xia et al. 2021)
AChE/Ag@Ti$_3$C$_2$T$_x$/Glassy carbon electrode	Malathion	DPV	0.01 pM–0.01 μM	0.00327 pM	(Jiang et al. 2018c)
CuO/LDG	Glucose	CV	1 μM–5 mM	0.1 μM	(Prabhakaran and Nayak 2020)
Ag-Pt nanorings decorated on rGO (AgPt NRs-rGO)	Carcinoembryonic antigen (CEA)	CV	5 fg/mL–50 ng/mL	1.43 fg/mL	(Wang et al. 2018a)

(Contd.)

Table 9.2 A summary of a few electrochemical-based biosensors (*Contd.*).

Materials used	Analyte	Measurement technique	Detection range	Limit of detection	References
$TiO_2/Ti_3C_2T_x$-$BiVO_4$	CD44 protein	CV	1×10^3–0.016 nM	7×10^{-8} nM	(Soomro et al. 2020)
Cu^{2+}loaded Zeolite A/N–GNS/GCE	Dopamine	SWV	4–100 nM	3.8 nM	(Hatefi-Mehrjardi let al. 2019)
ITO/Au/Ti_3C_2-tetrahedral DNA	cTnI	SWV	0.1×10^{-6}–1×10^{-3} nM	0.04×10^{-6} nM	(Mi et al. 2021)
CuONPs-modified PIGE	Riboflavin	SWV	3.13–56.3 nM	1.04 nM	(Sukumar et al. 2020)
Ap-BP NSs/TH/Cu-MOF/GCE	MicroRNA (miR3123)	SWV	2 pM–2 µM	0.3 pM	(Sun et al. 2020)
EBT MIP/Amine/PEDOT/LDG	Chloramphenicol	EIS	1 nM–10 mM	0.62 nM	(Cardoso et al. 2019)
rGO-AgNPs/Glassy carbon electrode	H_2O_2	CPA	0.002–20 mM	0.73 µM	(Salazar et al. 2019)
GOx/CNTs/$Ti_3C_2T_x$/PB/CFMs	Glucose	CA	10 µM–1.5 mM	0.33 µM	(Lei et al. 2019)
MoS_2	TNFα	Impedance	1–200 pg ml^{-1}	0.01 pg/ml	(Sri et al. 2022)

simple hydrothermal method and was followed by doping of Cu ions by chemical mixing method. The square wave voltammetry (SWV) used method showed good sensitivity and LOD of 3.8 nM. This work did not report cross-selectivity and long-term stability studies.

Recently, black phosphorous (BP) attracted considerable attention in the development of highly sensitive and reliable biosensor. However, the BP loses its stability under ambient conditions due to its reactive nature to water and humidity and this eventually degrades its electrochemical properties. Therefore, the functionalization strategy was used to increase its sensing performance. Another metal along with a thionine-doped 2D MOF surface was utilized to detect miR3123 in real samples (Sun et al. 2020). The MOF with a high surface area and available vacancies was very useful in sensing. The BPNSs/TH/Cu-MOF/glassy carbon electrode (GCE) was loaded with ferrocene (Fc)-labeled single-strand DNA aptamer and when it interacted with miR3123 directed Fc move from GCE, which decreased in redox peak current of Fc. The sensor exhibited good selectivity to miRNA-3123. Another laser-applied graphene used biosensor was reported to detect chloramphenicol (Cardoso et al. 2019). A molecular-imprint polymer was fabricated over the electrode surface using a direct electro-polymerization of eriochrome black T (EBT). The sensor showed a linear response to 1 nM–10 mM and LOD of 0.62 nM. However, the sensor revealed a good selectivity to other interfering antibiotics such as (OTC, AMC, and sulfadiazine). The carbon and metallic nanomaterials-based nanocomposites were also studied for the development of biosensors (Salazar et al. 2019). The increase in reduction current was observed for rGox/AgNPs nanocomposite against H_2O_2 measured by constant potential amperometry (CPA), suggesting the electrocatalytic action of the AgNPs and the amplification of the electron transmission kinetics attributed to graphene. The sensor presented good performance of 7 months long-term stability with a low loss and fast response time of 2 s, only.

9.3 FIELD-EFFECT TRANSISTORS-BASED BIOSENSORS

Nowadays, FETs are very promising due to their flexible sensing platform, versatile, portable, and low-cost approach. The device can act as a sensor and also as an amplifier. Along with that, the FETs have several advantages, such as high sensitivity, selectivity, no labeling, easy working process, less chemical consumption, and fast analysis, etc. Here, some FETs-based biosensors are discussed below and also a summary of a few sensors is shown in Table 9.3.

Table 9.3 Development of a few FETs-based biosensors

Materials used	Deposition method	Analyte	Detection range	Limit of detection	References
ZIF-67 derived porous Co_3O_4 solution-gated graphene	CVD	Glucose	10 nM–1 mM	100 nM	(Xiong et al. 2018)
Graphene	Exfoliation	DNA	1 pM–10^5 pM	10 fM	(Wang and Jia 2018)
Mo doped ZnO	Magnetron sputtering	Hepatitis B virus DNA	1 pM–10 μM	1 pM	(Shariati et al. 2022)
ZnO (NPs) doped MoS_2	Vapor-liquid-solid (VLS)	HBV DNA	0.5 pM–50 μM	1 fM	(Shariati et al. 2021)
CNTs	–	SARS-CoV-2 RNA	1 nM–10 fM	10 fM	(Thanihaichelvan et al. 2022)
Graphene oxide-graphene	–	SARS-CoV-2 proteins	10 fg/mL–100 pg/mL	~8 fg/mL	(Gao et al. 2022)
Electrical double layer (EDL)	–	SARS-CoV-2 nucleocapsid (N) protein	0.4–400 ng/mL	~3 pM	(Chen et al. 2022)
CNTs	Inkjet printing	SARS-CoV-2 spike protein (S1)	0.1 fg/mL–5.0 pg/mL	4.12 fg/mL	(Zamzami et al. 2022)
TiN	DC sputtering	cTn-I antibodies	0.01–100 ng/mL	0.01 mV/$pC_{cTn\text{-}I}$	(Pan et al. 2022)
Si	–	cTn-I antibodies	0–0.5 ng/mL	0.016 ng/mL	(Chang et al. 2020)
Graphene	CVD	Imatinib	0.1 pM to 10 μM	15.5 fM	(Xu et al. 2021a)
GOD-GA-Ni/Cu-MOFs	Hydrothermal method and drop casting	Glucose	0.001–20 mM	0.51 μM	(Wang et al. 2021c)
Fe_3O_4@AuNPs	–	Dopamine	1–120 μmol/L	3.3 mol/L	(Liu et al. 2021b)
Graphene	CVD	Glucose	1 μM–10 mM	200 nM	(Wang et al. 2021a)
InN	PECVD	HIV gp41 antibodies	0–400 ng/mL	2.5 pM	(Song et al. 2021)
Indium gallium zinc oxide	Radio frequency sputtering	miRNA-21	10^{-16} mol/L – 10^{-9} mol/L	19.8 mol/L	(Guo et al. 2022)

The solution-gated graphene transistors (SGGTs) are used as a simple and low-cost biosensor for the instantaneous sensing of glucose and uric acid in tears. Combined nanomaterials like Au gated through ZIF-67 derivative porous Co_3O_4 hollow nanopolyhedrons are adapted with GOx-CHIT and BSA-CHIT respectively (Xiong et al. 2018). The functionalized porous Co_3O_4 hollow nanopolyhedrons acted as part of the electrocatalyst and significantly enhanced the H_2O_2 reduction and improved the sensitivity. The glucose and uric acid device showed a LOD < 100 nM. Also, the functionalized SGGT devices detect the 323.2 ± 16.1 μM and 98.5 ± 16.3 μM glucose and uric acid in actual tears. However, the sensor showed good selectivity to glucose than uric acid and in the case of glucose sensing, the stability was studied for only 5 days. Another FETs-based biosensor was reported for DNA sensing by utilizing graphene-based solution-gated field effect transistors (G-SgFETs) (Wang and Jia 2018). The liquid exfoliated graphene (LEG) was fabricated and explored sensing due to its rich oxygen containing functional groups which may bind the biosensing probes. The double GA cross-linking together LEG film coating and biofunctionalization was applied. Due to the change in surface states, the surface become hydrophilic to hydrophobic which was important to bind the DNA molecules. The sensor showed a good response to the concentration of 1 pM to 10^5 pM with LOD of 10 fM. The cross-selectivity and stability of the sensor were not considered. Recently, another hepatitis B virus deoxyribonucleic acid (HBV DNA) sensor was developed using molybdenum doped ZnO nanowires (NWs) (Shariati et al. 2022). The material was prepared using a physical vapor deposition (PVD) method and modified with doping of molybdenum, which brought active sites to DNA binding and electric charge transmission. This device was verified by examining PCR-confirmed analytes like True Positive (TP), True Negative (TN), False Positive (FP), and False Negative (FN). The sensor showed a good sensing performance in between the concentration of 1 pM to 10 μM and achieved a LOD ~1 pM. Despite, large responsivity of the device, the stability and cross-selectivity were not studied.

The new type of coronavirus SARS-CoV-2 was also detected by using a FETs-based technique. The FET was modified with the GO/Gr heterojunction and explored to detect SARS-CoV-2 proteins (Gao et al. 2022). In the actual experiment, the SARS-CoV-2 spike protein solution was prepared in 10 fg/mL–100 pg/mL and detected by the FET device. With the addition of the concentration, the V_{Dirac} of GO/Gr FET was shifted to the left. The sensor exhibited a sensitivity of 12.8 mV/decade. However, a slight change in detection sensitivity because of the quality and process variation of the graphene in different devices was observed. Also, the biosensor was taking more time 20 min to detect 8 fg/mL SARS-Cov-2 protein, which might be time-consuming. Another

Saliva-used COVID-19 rapid antigen test of SARS-CoV-2 nucleocapsid protein EGFETs was also reported (Chen et al. 2022). It contained a portable smartphone-based reader interface as shown in Figure 9.2. The device performance was tested to SARS-CoV-2 N protein solution ranging from 0.4, 4, 40, to 400 ng/mL and achieved the LOD of 342.16 pg/mL. Despite its simple fabrication method, the time taken for the measurement was a little longer around 20 min, and the reversal time around <30 min, which might affect the rapid COVID-19 antigen tests.

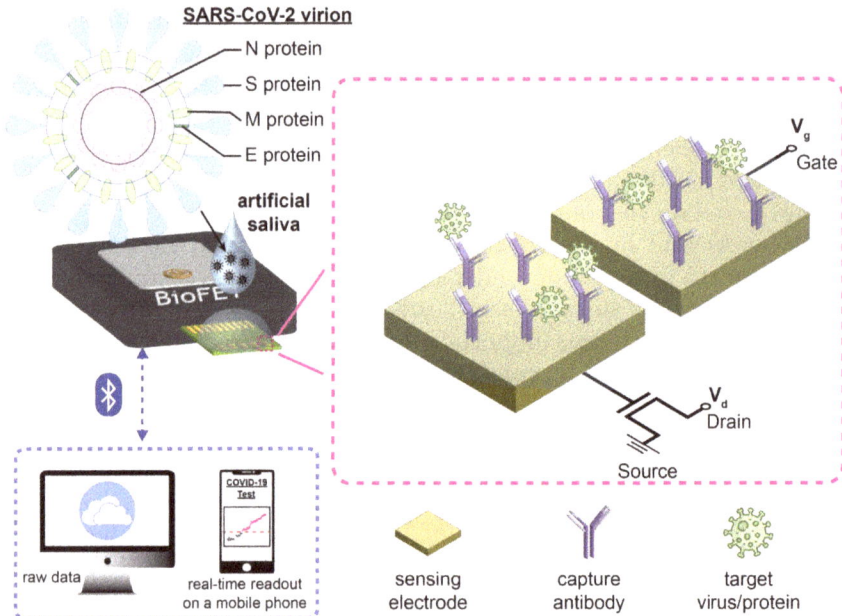

Figure 9.2 Saliva-used COVID-19 rapid antigen test of SARS-CoV-2 nucleo-capsid protein EGFETs sensor. [Reproduced with permission (Chen et al. 2022). Copyright 2022, Elsevier].

Recently, metallic nanoparticles doped MOF were also tested for glucose detection. The dual metal Ni/Cu-MOFs synthesized via a hydrothermal technique was coated by a drop cast to obtain the sensing surface to measure the glucose in 1 μM–20 mM (Wang et al. 2021c). As a result of the collaborative result of Ni and Cu ions in MOFs, the device (GOD-GA-Ni/Cu-MOFs-FET) exhibited good sensing with obtained sensitivity of 26.05 $\mu Acm^{-2}mM^{-1}$ to 1 to 100 μM with a LOD of 0.51 μM. Ni/Cu-MOFs acted as a p-type semiconductor on a gate voltage of −0.1 V and so, on the introduction of H^+, the strength of holes in Ni/Cu-MOFs reduced, subsequently decreasing the conductance and altering the corresponding current of FET with respect to the rise in glucose concentration ranging from minimum to maximum. While studying its stability over 6 days, the response of day 4 and day 6 decreased marginally than days 0 and 2 and thus, the sensor did not display long-term stability. The metal oxide and metal heterojunction also successfully utilized biochemical parameter detection. In-vivo measurement of dopamine in fish brains was detected by a Fe_3O_4@ AuNPs based EN-FETs geometry (Liu et al. 2021b). By varying the space between the magnet and the device, the response of this device was perfectly tuned. The sensor exhibited a significant response to dopamine alone as that of others. The chip was used for dopamine sensing in the crucian fish brain. However, it should be noted that the sensor output may get affected by external fields as well. The sensitive materials of tri-layer of SiOx/SiNx/InN thin film self-rolled into microtubes were used for HIV gp41 antibodies sensing (Song et al. 2021). InN exhibited a robust, inherent, and steady electron buildup ($\sim10^{13}$ cm^{-2}), and therefore, provided a good performance to HIV gp41 antibodies concentration. The inside surface of InN microtube was concave, and therefore, effectively in contact with the analyte solution enhanced the sensitivity. This device showed a LOD of around 0.1 ng/mL, considerably lower than standard ELISA. This study did not report the sensor selectivity and stability.

9.4 WHISPERING GALLERY MODE (WGM) RESONATORS-BASED BIOSENSORS

Whispering gallery mode (WGM) resonators are very promising in biochemical sensing. Generally, WGM consists of a light entrapped in a circle-shaped resonating structure, acting as an optical cavity. When the interference condition gets satisfied, then coupling resonance modes are only allowable and identified as fine pikes. In presence of external parameters (like a biological entity) there is a spectral shift which occurs in the spectrum which directly depends on the concentration of the surrounding parameter. The WGM-based biosensors are attractive due to their mesmerizing properties such as small size, high sensitivity, fast response, label-free, and on-time detection. The following are some examples discussing the importance along with the limitations of WGMs in biochemical applications. A summary of WGM resonators-based biosensors is displayed in Table 9.4.

A self-referenced biosensor containing two nearly the same sized dye-doped polystyrene microspheres which were located on nearby holes at the tip of a microstructure optical fiber and utilized for the detection of Neutravidin in undiluted, immunoglobulin-deprived human serum samples (Reynolds et al. 2016). Among these two microspheres, one microsphere acted as a reference which was used to compensate alterations in the nearby surrounding (such as refractive index and temperature), whereas the other microsphere was specifically used for the detection of the analyte. The increase in the surface density was observed in presence of a serum sample. But this study did not report the selectivity with other biological samples. The hemoglobin (HB) molecules in blood samples was quantified by using a WGM ring resonator which was coated by a sensitive layer deposited on the inside of it (Ajad et al. 2021). The numerical analysis showed that the wavelength shift will occur because of modification in the refractive index of HB. This device showed a wavelength shift of 7.5 to an RI variation of 1.32919–1.34995 which was analogous to HB concentration. The optimized geometry showed that a ring resonator used sensor exhibited a response of 361.3 nm/RIU with Q = 1143. Despite, its higher sensitivity, the sensors' selectivity to other interfering RI and stability was not analyzed. The Antigen-Antibody interactions were studied based on Protein G layered microsensors which depended upon the WGM (Álvarez Freile et al. 2021). The fluorescence spherelike microbead material with a diameter ~10 μm was used to find the interfaces among therapeutic monoclonal antibodies (mAbs) and proteins by cancer cells. The sensor time was very short in seconds, and therefore, could be more useful than the SPR method. The round spherical resonator modified with GOx and AuNPs was used (shown in Figure 9.3) for the detection of glucose (Brice et al. 2020). The enzyme oxidized glucose-by glucose oxidase (GOx) with gold nanoparticles overlayed to a WGM-resonating structure. Au NPs were coated by a dip coating method. The sensing performance of WGMRs/Au-NPs/GOx device and WGMRs/GOx was compared and it was observed that the former showed a higher sensitivity of 0.294 MHz/s and rising time in 60–120 s. The current sensing setup can also be used to measure glucose in a watery medium.

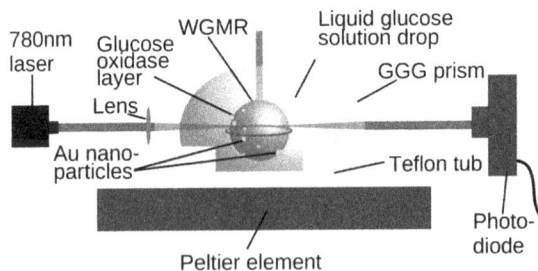

Figure 9.3 Illustration of an experimental setup based on WGM modified with GOx and AuNPs. The system consists of a GGG prism, lens and mirrors, WGM-resonator mounting, and a Peltier element for temperature maintenance. [Reproduced with permission (Brice et al. 2020). Copyright 2020, Elsevier].

Another biosensor-based on a 2-D material was also fabricated. In this work, SiO_2 dielectric microspheres modulated WGM characteristics which located on the top of Au NPs/MoS_2 nanosheets were used for quantification of the K-RAS gene (Wang et al. 2021b). The Au NPs exhibited reflective properties, and therefore, acted as antireflective layers. The plasmonic structures efficiently enhanced the localization of the electric field into the sensing area. The sensing bandwidth of the device was from 1 fM to 1 nM and LOD of 0.3 fM. However, the device sensing characteristics with interfering parameters need to be discussed. The WGM in liquid crystallite micro-drop based structure was reported for the detection of real-time, quantitative, and fast detection of urea (Duan et al. 2020b). The WGM response of acid-doped 5CB microdroplets was studied at different pH values. The sensor was performed in $10~mM^{-1}$ mM and obtained a responsivity of 8.1 nm/mM for urea molecules. The response time was in the range of 1–4 min.

Table 9.4 A summary of a few Whispering gallery mode (WGM) resonators-based biosensors

Type of WGM resonator	Analyte	Sensor response	References
Microspheres	Neutravidin	Highly selective	(Reynolds et al. 2016)
Microbottle resonator	BSA	Noise equal detection limit: 10 fg/ml	(Ghali et al. 2016)
SiO_2 WGM ring resonator	Hemoglobin	Sensitivity: 361.3 nm/RIU	(Ajad et al. 2021)
Polystyrene microbeads coated with protein G	Protein G	Wavelength shift: $\Delta\lambda$ = 93 pm	(Álvarez Freile et al. 2021)
Round sphere resonator	Glucose	Frequency shift: 0.294 MHz/s	(Brice et al. 2020)
Ni doped MoS_2 QDs coated microspheres	K-RAS gene	Detection limit as 0.3 fM	(Wang et al. 2021b)
Liquid crystal biosensor based on whispering gallery mode lasing	Fenobucarb and dimethoate	Limit of detection achieved < 0.1 pg/mL for fenobucarb and 1 pg/mL for dimethoate	(Duan et al. 2020a)
Micro resonator	R antigen	Wavelength shift: 15 pm; LOD: 0.124 μg/mL	(Wu et al. 2021)
Spherical WGM lasing-based LC sensor	Urea	Sensitivity: 8.1 nm/mM; LOD: 0.1 mM	(Duan et al. 2020b)

9.5 OPTICAL FIBER-BASED BIOSENSORS

Since the last decade, optical fiber has been tremendously explored for biosensing applications due to its various merits such as cost-effectiveness, small-size, flexibility, robustness, no electromagnetic interference, chemical inertness, lightweight, remote and multiplexed detection capability, respectively. Figure 9.4 shows some optical fiber designs typically used in biosensing. The summarized development of some biosensors-based on optical fiber is tabulated in Table 9.5.

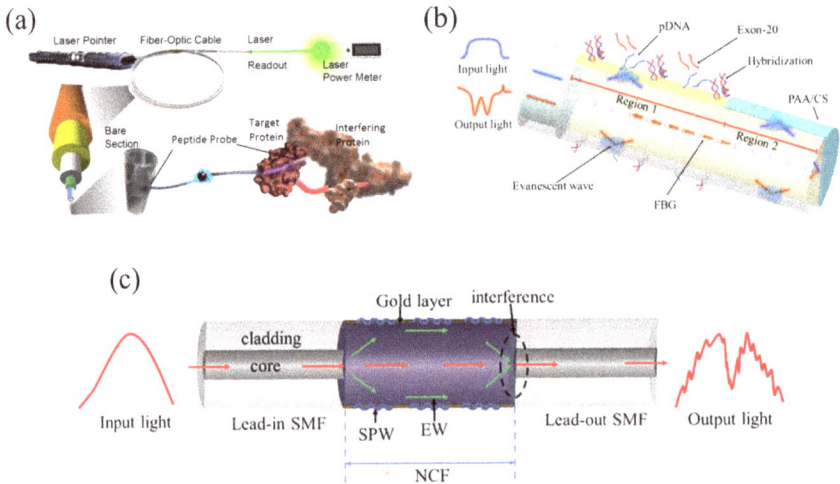

Figure 9.4 (a) Optical fiber-based proposed method to detect the main protease of SARS-Cov-2, [Reproduced with permission (Wang et al. 2021d). Copyright 2021, American Chemical Society]. (b) Sectional view of the optical fiber-based sensor consisting pDNA (probe DNA sequence), PAA/CS (poly (acrylic acid)/chitosan), Exon-20 (exon 20 fragment of EGFR gene), [Reproduced with permission (Li et al. 2022b). Copyright 2022, Elsevier]. and (c) Schematic diagram of the SPR and MZI hybrid optical fiber DNA sensor. [Reproduced with permission (Gong et al. 2021). Copyright 2021, Elsevier].

Numerous gratings-based structures have been used for the detection of biochemical parameters. In this, an enzymatic graphene oxide (GO)-functionalized tilted fiber grating structure was studied for glucose sensing (Jiang et al. 2018a). The tilted grating structure was used because it offered a high sensitivity to local ambiance RI change as the evanescent field strongly interacted with the high-order cladding mode of the fiber. When the particular catalysis of GOD on the glucose occurred, the corresponding change in the RI led to resonant wavelength shifts in the interference of the TFG. The sensor showed that during low glucose concentration from 0–8 mM, the sensitivity in terms

Table 9.5 A summary of a few some optics fiber-based biosensors

Materials used	Fiber type	Deposition method	Analyte	Detection range	LOD	References
GO/GOD	LPG (TFG)	Dip coating	Glucose	0–8 mM	–	(Jiang et al. 2018a)
4-ANMP/PVC	FPI (SMF)	Dip coating	Glucose	1 µM-1 M	–	(Khan et al. 2018)
XO/Ta$_2$O$_5$ NPs	SPR (UMMF)	Electrospinning	Xanthine	0–3 µM	0.0656 µM	(Kant et al. 2018)
PAA/CS	FBG	Layer by layer	pDNA	0–100 nM	13.5 nM	(Li et al. 2022b)
PLL/ssDNA	Microfiber	–	DNA	0.1 nM-1 µM	75 pM	(Gao et al. 2017)
APTES+pDNA	C-type fiber +SMF Fabry-Perot interference (FPI)	Chemical modification	DNA hybridization	–	68 nM	(Li et al. 2022a)
ssDNA	SPR (SMF)	Self-assembled monolayer	DNA	10 fM–100 nM	10 fM	(Kaye et al. 2017)
3-mercaptopropyltrimethylsilane (MPTMS) -Au-zearalenone nucleic acid aptamer	End facet coated	Chemical modification	Zearalenone	1–480 ng/mL	0.102 ng/mL	(Xu et al. 2021b)
PS-*b*-PV4P-templated citrate-AuNR film	Uncladded optical fiber LSPR	Dip coating method	Human IgG	0–667 nM	0.6 nM	(Lu et al. 2021)
Graphene oxide	Tilted fiber grating	Chemical modification	Hemoglobin	0.1 mg/ml – 1.0 mg/ml	0.4 mg/ml	(Sun et al. 2022)

(Contd.)

Table 9.5 A summary of a few some optics fiber-based biosensors (*Contd.*)

Materials used	Fiber type	Deposition method	Analyte	Detection range	LOD	References
Uric acid-uricase/Au/ SiO₂/optical fiber	Unclad optical fiber	Chemical modification	Uric acid	0.05 mM– 1 mM	0.02 mM	(Jain et al. 2022)
GO/Au	Multimode fiber	Chemical modification	Chiral amino acids (AAs)	5×10^{-4} mM– 30 mM	1.09×10^{-9} mM	(Zhou et al. 2022)
APTES-functionalized and laccase immobilized CDD-CDs	Tapered Plastic cladded silica fiber	Dip coating method	Dopamine	0–10 μM	46.4 nM	(Sangubotla and Kim 2021)
Anti-NT-proBNP MAbs	LPG (TFG)	Chemical modification	NT-proBNP	0.5–1000 ng/mL	0.5 ng/mL	(Luo et al. 2017)
AbCK7	SPR (TFBG)	Chemical modification	Cytokeratin 7 peptide	1–100 ng/mL	0.4 nM	(Ribaut et al. 2017)
PLL/anti-cTn-I	FBG (MF)	SA	cTn-I	0.03–100 ng/mL	10.8 pg/mL	(Liu et al. 2018)
Anti-PSA solution	High-order-diffraction long period grating (HOD-LPG)	Functionalization technique	PSA	5 to 500 ng/mL	9.9 ng/mL	(Xiao et al. 2020)

of ~0.24 nm/mM was obtained. However, the sensor selectivity and long-term stability need to be assessed for practical applications. The effect of humidity and temperature was also not discussed. Another glucose sensor was fabricated by using an Fabry-Perot interferometer approach. In this work, different solvatochromic dyes (like Nile red, rhodamine-B, and 4-amino-N-methylphthalimide) were separately combined into the polymeric structure and employed for glucose sensing (Khan et al. 2018). The Au layer was used over the material surface for better reflection. The response and recovery times of these sensing geometries were fast at around 8 s and 9 s and showed good sensitivity of 3.25 nm/mM in 1 µM–1 M. The sensing configuration was simple to construct and operate. The metal oxides-based SPR sensor fabricated by using a xanthine oxidase functionalized Ta_2O_5 nanostructures along with Ag coating on clad removed fiber was reported for xanthine detection (Kant et al. 2018). The change in the dielectric constant upon exposure to xanthine concentrations led to a shift in resonance wavelength. The sensor was adjusted over pH and XO captured in Ta_2O_5. The sensor showed a response of ~26.2 nm/µM to 0–3 µM and LOD ~0.0127 µM. However, after each measurement, the sensor needed to be washed thoroughly and dried in the air, which might be time-consuming. A Sagnac interferometer-based DNA biosensor was also fabricated. The polarimetric interference of a high-birefringence (Hi-Bi) fiber was effectively used to detect DNA molecules (Gao et al. 2017). The fiber was tapered up to the waist diameter of 5.2 µm, so that the light effectively interacted with the DNA. When there was a little change in the concentration of DNA, this led to an observed resonance shifts. The probe detected a DNA concentration of 100 pM–1 µM with a LOD of 75 pM. In the selectivity study, the response of the sensor was only tested for a few interfering molecules.

A phase-shifted microfiber Bragg grating probe was utilized for the detection of cardiac troponin I (cTn-I) (Liu et al. 2018). The layer-by-layer self-assembly method to achieve better functionalization was used. In the specificity study, only two types of non-specific proteins–ALB and IgG were considered. Also, the sensor was tested to a long range of cTn-I concentration of 0.03–100 ng/mL, but it showed a linear response to 0.1–10 ng/mL range only. The prostate specific antigen (PSA) detection was performed by using a high-order-diffraction long-period grating (HOD-LPG) (Xiao et al. 2020). The HOD-LPG was biofunctionalized via several biochemical methods in which the surface of the tapered part was coated by PSA antibody. When the PSA protein of concentration of 500 ng/ml interacted with the binding surface, the total spectral shift of 0.37 nm was observed. The sensor temperature and bending response were also cross-verified. However, the sensor did not show a linear response in lower concentration range.

9.6   COLORIMETRIC AND FLUORESCENCE-BASED BIOSENSORS

The colorimetric detection by phage-enzyme-linked immunosorbent assay (P-ELISA) and two phage time-resolved fluoroimmunoassays (P-TRFIAs) was performed (Du et al. 2020). The half-maximum inhibition concentrations (IC50) of the P-ELISA, P-TRFIA-1, and P-TRFIA-2 were 0.067 ng/mL, 0.085 ng/mL, and 0.056 ng/mL, consecutively. The sensor showed a good performance. However, the device also displayed a lengthy sensing period, interference with imidaclothiz and clothianidin, etc. The carbendazim (CBZ) colorimetric detection in spiky water solution by definite aptamers of carbendazim (CBZ), Au NPs and cation poly-diallyldimethylammonium chloride (PDDA) was reported (Wang et al. 2020). It was observed that because intake of aptamer, the PDDA did not interact with aptamer and started inducing aggregation of AuNPs, and thus, caused a changed in color from red to blue. This showed a linear response to 2.2 to 500 nM with a LOD of 2.2 nM. The study showed that the sensor presented a good response and less sensing time. However, this device was not assessed for food. This method was only applied to water-based samples. The molecularly imprinted polymer was also used effectively to detect the biochemical parameter through the colorimetric method. A molecular imprinted polymer was prepared for the identification of 3-Phenoxybenzaldehyde and utilized for the detection of 3-Phenoxybenzaldehyde (3-PBD) the metabolite of pyrethroid pesticides (Ye et al. 2018). The sensor showed a good linear response to 0.1–1 µg/mL and LOD ~0.052 µg/mL. The sensor displayed advantages like large response and speedy while, the disadvantages was that it was not evaluated to different oxidative components. A biochemical parameter by using metal nanoparticles was also detected. The AuNPs were altered by monoclonal antibodies and 6-carboxyfluorescein marked single-stranded thiol-oligonucleotides (6-FAM-SH-ssDNAs), where the intensity was quenched by gold NPs. It observed that the sensor intensity was inversely related to the concentration (Zhang et al. 2018). The sensor showed a linear response to 0.01–20 µg/L with LOD of 6 ng/L. This sensor offered high sensitivity and selectivity. However, there are certain limitations like a continue, a lengthy detecting period of 2.5 h, and therefore, making it complicated for multi-modification of AuNPs. A summary of colorimetric and fluorescence-based biosensors is tabulated in Table 9.6.

Table 9.6 A summary of a few colorimetric and fluorescence-based biosensors

Detection technique	Bioreceptor element	Analyte	Detection range (μM)	Limit of detection (μM)	References
Colorimetric	Antibody	Imidacloprid	9.39×10^{-5}–1.56×10^{-3}	9.39×10^{-5}	(Du et al. 2020)
Colorimetric	Aptamer	Carbendazim	2.20×10^{-3}–5.00×10^{-1}	2.20×10^{-3}	(Wang et al. 2020)
Colorimetric	MIP	3-Phenoxybenzaldehyde	5.04×10^{-1}–5.04	2.62×10^{-1}	(Ye et al. 2018)
Colourimetry	GO-based nanoprobe	mRNA	1–400 nM	0.26 nM	(Jiang et al. 2018b)
Colourimetry	DNA-AgNCs-GO	CA 125	2 ng/mL–6.7 μg/mL	1.26 ng/mL	(Wang et al. 2018b)
Colourimetry	GO	CEA	28.5 fg/mL	–	(Xu et al. 2018)
Colourimetry	H-GNs	Telomerase	100–2300 cells/mL	60 cells/mL	(Xu et al. 2017)
Fluorescence	Antibody	Triazophos	3.19×10^{-5}–6.38×10^{-2}	1.92×10^{-5}	(Zhang et al. 2018)
Fluorescence	Aptamer	Isocarbophos	1.00×10^{-2}–5.00×10^{-1}	1.00×10^{-2}	(Li et al. 2018)
Fluorescence	MIP	Pyrethroids	1.00–8.00	1.30×10^{-1}	(Li et al. 2017)
Fluorescence	MIP	Atrazine	2.32–1.85×10^{2}	8.60×10^{-1}	(Liu et al. 2016)
Chemiluminescence	Aptamer	Acetamiprid	8.00×10^{-4}–6.30×10^{-1}	6.20×10^{-5}	(Qi et al. 2016)

9.7 CONCLUSION AND OUTLOOK

In recent years, biosensors have made rapid growth in various fields including healthcare and diagnostics, medical, and agriculture. In this chapter, various detection systems for biosensing applications have been reviewed. It was observed that the different biosensors are prepared based on the functions and diagnosis. Various materials including MOXs, metal coatings, and polymeric materials are applied for the detection of biochemical species. These materials offered great scope for functionalization, long-term stability, and flexibility. However, much research is still required for optimization and making them available for real-world clinical applications.

The following are key points regarding techniques used and characteristics of biosensing.

1. Sensitivity is a very important parameter in biosensing. Generally, aptamers offered a high response owing to their small size. The nucleic acids exhibited poor sensitivity because of steric interruption. In the case of MIPs, due to flexibility in structure, the sensitivity and stability can be enhanced.

2. Selectivity is a crucial parameter in biosensing. It measures the ability to detect the minimum concentration of the analyte among the various interfering components. The selectivity is hampered in the case of nucleoid acid and aptamers and identification components. The electrostatic interfaces get reduced because of peptide nucleic acids. MIPs also showed poor selectivity.

3. It detected that the change in the electrode material, experimental parameters, and membrane films greatly improved the detection range of a biosensor.

4. At present, the DNA-based sensing probes are attractive due to their ability of sequencing info. However, in the case of point alterations, it fails.

5. Glucose detection has been done by most biosensing methods. A lot of commercial glucose biosensors also have been reported.

6. Graphene-based materials have been widely used in biosensor development because of great surface occupancy and high conductivity. However, the cytotoxicity needs to be studied and thus, its biosafety must be ensured.

7. Many metal-coated nanocomposites-based biosensors have been reported so far. It provided a high response, high selectivity, and good stability. These techniques have great potential to address the problems. Still, they are at an early stage of development, and therefore, need to be studied more.

8. The processes like repeated filtration and purification techniques may concentrate on the target analyte, and therefore, need to be analyzed.

9. At present, rapid testing kits are attracting great attention because of their portability, simple fabrication, low cost, and effectiveness in sensing. However, reliability and sensitivity need to improve.

10. There is scope for the development of a disposable and biodegradable materials-based biosensor development.

11. Smartphone-based hand-held devices are of growing interest due to their miniaturization, ease in use, and real-time detection.

12. Wearable biosensors have been of immense interest due to their fast detection, ease of use, compact size, real-time monitoring, and low power consumption. The future trend includes improving detection accuracy and flexibility.

13. Optimizing parameters for biosensing developments is very crucial for better understanding the information and reliability in diagnosis.

REFERENCES

Ajad, Abul Kalam, Md Jahirul Islam, Md Rejvi Kaysir and Javid Atai. 2021. Highly sensitive bio sensor based on WGM ring resonator for hemoglobin detection in blood samples. Optik 226: 166009. doi:https://doi.org/10.1016/j.ijleo.2020.166009.

Álvarez Freile, Jimena, G. Choukrani, Kerstin Zimmermann, Edwin Bremer and Lars Dähne. 2021. Whispering gallery modes-based biosensors for real-time monitoring and binding characterization of antibody-based cancer immunotherapeutics. Sensors and Actuators B: Chemical 346: 130512. doi:https://doi.org/10.1016/j.snb.2021.130512.

Akshaya, K.B., Vinay S. Bhat, Anitha Varghese, Louis George and G. Hegde. 2019. Non-enzymatic electrochemical determination of progesterone using carbon nanospheres from onion peels coated on carbon fiber paper. Journal of The Electrochemical Society 166(13): B1097–1106. The Electrochemical Society. doi:10.1149/2.0251913jes.

Brice, Inga, Karlis Grundsteins, Aigars Atvars, Janis Alnis, Roman Viter and Arunas Ramanavicius. 2020. Whispering gallery mode resonator and glucose oxidase based glucose biosensor. Sensors and Actuators B: Chemical 318: 128004. doi:https://doi.org/10.1016/j.snb.2020.128004.

Cardoso, Ana R., Ana C. Marques, Lídia Santos, Alexandre F. Carvalho, Florinda M. Costa, Rodrigo Martins, M. Goreti F. Sales, et al. 2019. Molecularly-imprinted chloramphenicol sensor with laser-induced graphene electrodes. Biosensors and Bioelectronics 124–125: 167–175. doi:https://doi.org/10.1016/j.bios.2018.10.015.

Chaiyo, S., E. Mehmeti, W. Siangproh, Thai Long Hoang, H. Phong Nguyen, Orawon Chailapakul and Kurt Kalcher. 2018. Non-enzymatic electrochemical detection of glucose with a disposable paper-based sensor using a cobalt phthalocyanine–ionic liquid–graphene composite. Biosensors and Bioelectronics 102: 113–120. doi:https://doi.org/10.1016/j.bios.2017.11.015.

Chang, Shih-Mein, Sathyadevi Palanisamy, Tung-Ho Wu, Chiao-Yun Chen, Kai-Hung Cheng, Chen-Yi Lee, Shyng-Shiou F. Yuan, et al. 2020. Utilization of silicon nanowire field-effect transistors for the detection of a cardiac biomarker, cardiac troponin I and their applications involving animal models. Scientific Reports 10(1): 22027. doi:10.1038/s41598-020-78829-7.

Chen, Pin-Hsuan, Chih-Cheng Huang, Chia-Che Wu, Po-Hsuan Chen, A. Tripathi and Yu-Lin Wang. 2022. Saliva-based COVID-19 detection: a rapid antigen test of SARS-CoV-2 nucleocapsid protein using an electrical-double-layer gated field-effect transistor-based biosensing system. Sensors and Actuators B: Chemical 357: 131415. doi:https://doi.org/10.1016/j.snb.2022.131415.

Du, Mei, Qian Yang, Weimei Liu, Yuan Ding, He Chen, Xiude Hua and Minghua Wang. 2020. Development of immunoassays with high sensitivity for detecting imidacloprid in environment and agro-products using phage-borne peptides. Science of The Total Environment 723: 137909. doi:https://doi.org/10.1016/j.scitotenv.2020.137909.

Duan, Rui, Xiaolei Hao, Yanzeng Li and Hanyang Li. 2020a. Detection of acetylcholinesterase and its inhibitors by liquid crystal biosensor based on whispering gallery mode. Sensors and Actuators B: Chemical 308: 127672. doi:https://doi.org/10.1016/j.snb.2020.127672.

Duan, Rui, Yanzeng Li, Bojian Shi, Hanyang Li and Jun Yang. 2020b. Real-time, quantitative and sensitive detection of urea by whispering gallery mode lasing in liquid crystal microdroplet. Talanta 209: 120513. doi:https://doi.org/10.1016/j.talanta.2019.120513.

Gao, Shuai, Li-Peng Sun, Jie Li, Long Jin, Yang Ran, Yunyun Huang and Bai-Ou Guan. 2017. High-sensitivity DNA biosensor based on microfiber sagnac interferometer. Optics Express 25(12): 13305–13313. OSA. doi:10.1364/OE.25.013305.

Gao, J.i, C. Wang, Yujin Chu, Y. Han, Y. Gao, Y. Wang, C. Wang, et al. 2022. Graphene oxide-graphene Van der Waals heterostructure transistor biosensor for SARS-CoV-2 protein detection. Talanta 240: 123197. doi:https://doi.org/10.1016/j.talanta.2021.123197.

Ghali, Hala, Hicham Chibli, J.L. Nadeau, Pablo Bianucci and Yves-Alain Peter. 2016. Real-time detection of staphylococcus aureus using whispering gallery mode optical microdisks. Biosensors 6(2): 20. https://doi.org/10.3390/bios6020020.

Gong, Pengqi, Yiming Wang, Xue Zhou, Shankun Wang, Yanan Zhang, Yong Zhao, Linh Viet Nguyen, et al. 2021. In situ temperature-compensated DNA hybridization detection using a dual-channel optical fiber sensor. Analytical Chemistry 93(30): 10561–10567. American Chemical Society. doi:10.1021/acs.analchem.1c01660.

Goodchild, Sarah A., Lee J. Hubble, Rupesh K. Mishra, Zhanhong Li, K. Yugender Goud, Abbas Barfidokht, Rushabh Shah, et al. 2019. Ionic liquid-modified disposable electrochemical sensor strip for analysis of fentanyl. Analytical Chemistry 91(5): 3747–3753. American Chemical Society. doi:10.1021/acs.analchem.9b00176.

Guo, Jing, Ruichen Shen, Xuejie Shen, Bo Zeng, Nianjun Yang, Huageng Liang, Yanbing Yang, et al. 2022. Construction of high stability indium gallium zinc oxide transistor biosensors for reliable detection of bladder cancer-associated MicroRNA. Chinese Chemical Letters 33(2): 979–982. doi:https://doi.org/10.1016/j.cclet.2021.07.048.

Haritha, V.S., S.R. Sarath Kumar and R.B. Rakhi. 2022. Amperometric cholesterol biosensor based on cholesterol oxidase and Pt-Au/MWNTs modified glassy carbon electrode. Materials Today: Proceedings 50: 34–39. doi:https://doi.org/10.1016/j.matpr.2021.03.128.

Hatefi-Mehrjardi, Abdolhamid, Amirkhosro Beheshti-Marnani and Nahid Askari. 2019. Cu^{+2} loaded 'zeolite a'/nitrogen-doped graphene as a novel hybrid for simultaneous voltammetry determination of carbamazepine and dopamine. Materials Chemistry and Physics 225: 137–144. doi:https://doi.org/10.1016/j.matchemphys.2018.12.073.

Jain, Surbhi, Ayushi Paliwal, Vinay Gupta and Monika Tomar. 2022. Smartphone integrated handheld long range surface plasmon resonance based fiber-optic biosensor with tunable SiO_2 sensing matrix. Biosensors and Bioelectronics 201: 113919. doi:https://doi.org/10.1016/j.bios.2021.113919.

Jalalvand, Ali R., Ali Haseli, Farshad Farzadfar and Hector C. Goicoechea. 2019. Fabrication of a novel biosensor for biosensing of bisphenol a and detection of its damage to DNA. Talanta 201: 350–357. doi:https://doi.org/10.1016/j.talanta.2019.04.037.

Jiang, Biqiang, Kaiming Zhou, Changle Wang, Qizhen Sun, Guolu Yin, Zhijun Tai, Karen Wilson, et al. 2018a. Label-free glucose biosensor based on enzymatic graphene oxide-functionalized tilted fiber grating. Sensors and Actuators B: Chemical 254: 1033–1039. doi:https://doi.org/10.1016/j.snb.2017.07.109.

Jiang, Hongyan, Fu-Rong Li, Wei Li, Xiaodong Lu and Kai Ling. 2018b. Multiplexed determination of intracellular messenger RNA by using a graphene oxide nanoprobe modified with target-recognizing fluorescent oligonucleotides. Microchimica Acta 185(12): 552. doi:10.1007/s00604-018-3090-1.

Jiang, Y., X. Zhang, L. Pei, S. Yue, L. Ma, L. Zhou, et al. 2018c. Silver nanoparticles modified two-dimensional transition metal carbides as nanocarriers to fabricate acetycholinesterase-based electrochemical biosensor. Chemical Engineering Journal: 339: 547–556. doi:10.1016/j.cej.2018.01.111..

Kant, Ravi, Rana Tabassum and Banshi D. Gupta. 2018. Xanthine oxidase functionalized Ta_2O_5 nanostructures as a novel scaffold for highly sensitive SPR based fiber optic xanthine sensor. Biosensors and Bioelectronics 99: 637–645. doi:https://doi.org/10.1016/j.bios.2017.08.040.

Karaman, Ceren, Onur Karaman, Necip Atar and Mehmet Lütfi Yola. 2021. Electrochemical immunosensor development based on core-shell high-crystalline graphitic carbon nitride@carbon dots and $Cd_{0.5}Zn_{0.5}S/d-Ti_3C_2T_x$ MXene composite for heart-type fatty acid–binding protein detection. Microchimica Acta 188(6): 182. doi:10.1007/s00604-021-04838-6.

Kaye, Savannah, Zheng Zeng, Mollye Sanders, Krishnan Chittur, P.M. Koelle, Robert Lindquist, Upender Manne, et al. 2017. Label-free detection of DNA hybridization with a compact LSPR-based fiber-optic sensor. Analyst 142(11): 1974–1981. The Royal Society of Chemistry. doi:10.1039/C7AN00249A.

Khan, M.R.R., A.V. Watekar and S. Kang. 2018. Fiber-optic biosensor to detect pH and glucose. IEEE Sensors Journal 18(4): 1528–1538. doi:10.1109/JSEN.2017.2786279.

Kim, Byungjoo, Hayoung Jeong, Yong Soo Lee, Seongjin Hong and Kyunghwan Oh. 2021. Spatially selective DNA deposition on the fiber core by optically trapping an aqueous droplet and its application for ultra-compact DNA fabry-perot temperature sensor. Sensors and Actuators Reports 3: 100038. doi:https://doi.org/10.1016/j.snr.2021.100038.

Lei, Yongjiu, Wenli Zhao, Yizhou Zhang, Qiu Jiang, Jr-Hau He, Antje J. Baeumner, Otto S. Wolfbeis, et al. 2019. A MXene-based wearable biosensor system for high-performance in vitro perspiration analysis. Small 15(19): 1901190. John Wiley & Sons, Ltd. doi:https://doi.org/10.1002/smll.201901190.

Li, Hongji, Xiao Wei, Yeqing Xu, Kai Lu, Yufeng Zhang, Yongsheng Yan and Chunxiang Li. 2017. A thin shell and 'sunny shape' molecular imprinted fluorescence sensor in selective detection of trace level pesticides in river. Journal of Alloys and Compounds 705: 524–532. doi:https://doi.org/10.1016/j.jallcom.2016.12.239.

Li, Xiaotong, Xiaomin Tang, Xiaojie Chen, Baohan Qu and Lihua Lu. 2018. Label-free and enzyme-free fluorescent isocarbophos aptasensor based on MWCNTs and G-Quadruplex. Talanta 188: 232–237. doi:https://doi.org/10.1016/j.talanta.2018.05.092.

Li, Fei, Xuegang Li, Xue Zhou, Pengqi Gong, Yanan Zhang, Yong Zhao, Linh Viet Nguyen, et al. 2022a. Plug-in label-free optical fiber DNA hybridization sensor based on C-type fiber vernier effect. Sensors and Actuators B: Chemical 354: 131212. doi:https://doi.org/10.1016/j.snb.2021.131212.

Li, Xuegang, Pengqi Gong, Qiming Zhao, Xue Zhou, Yanan Zhang and Yong Zhao. 2022b. Plug-in optical fiber SPR biosensor for lung cancer gene detection with temperature and pH compensation. Sensors and Actuators B: Chemical 359: 131596. doi:https://doi.org/10.1016/j.snb.2022.131596.

Liu, Guangyang, Tengfei Li, Xin Yang, Yongxin She, Miao Wang, Jing Wang, Min Zhang, et al. 2016. Competitive fluorescence assay for specific recognition of atrazine by magnetic molecularly imprinted polymer based on Fe_3O_4-chitosan. Carbohydrate Polymers 137: 75–81. doi:https://doi.org/10.1016/j.carbpol.2015.10.062.

Liu, Tong, Li-Li Liang, Peng Xiao, Li-Peng Sun, Yun-Yun Huang, Yang Ran, Long Jin, et al. 2018. A label-free cardiac biomarker immunosensor based on phase-shifted microfiber bragg grating. Biosensors and Bioelectronics 100: 155–160. doi:https://doi.org/10.1016/j.bios.2017.08.061.

Liu, Jingsi, Yuxia Fan, Gaole Chen and Yuan Liu. 2021a. Highly sensitive glutamate biosensor based on platinum nanoparticles decorated MXene-$Ti_3C_2T_x$ for L-glutamate determination in foodstuffs. LWT 148: 111748. doi:https://doi.org/10.1016/j.lwt.2021.111748.

Liu, Na, Xueping Xiang, Lei Fu, Qiang Cao, Rong Huang, Huan Liu, Gang Han, et al. 2021b. Regenerative field effect transistor biosensor for in vivo monitoring of dopamine in fish brains. Biosensors and Bioelectronics 188: 113340. doi:https://doi.org/10.1016/j.bios.2021.113340.

López-Fernández, Ester, Jorge Gil-Rostra, Juan P. Espinós, Ramon Gonzalez, Francisco Yubero, Antonio de Lucas-Consuegra and Agustín R. González-Elipe. 2020. Robust label-free $Cu_xCo_yO_z$ electrochemical sensors for hexose detection during fermentation process monitoring. Sensors and Actuators B: Chemical 304: 127360. doi:https://doi.org/10.1016/j.snb.2019.127360.

Lu, Mengdi, Hu Zhu, Ming Lin, Fang Wang, Long Hong, Jean-Francois Masson and Wei Peng. 2021. Comparative study of block copolymer-templated localized surface plasmon resonance optical fiber biosensors: CTAB or citrate-stabilized gold nanorods. Sensors and Actuators B: Chemical 329: 129094. doi:https://doi.org/10.1016/j.snb.2020.129094.

Luo, Binbin, Shengxi Wu, Zhonghao Zhang, Wengen Zou, Shenghui Shi, Mingfu Zhao, Nianbing Zhong, et al. 2017. Human heart failure biomarker immunosensor based on excessively tilted fiber gratings. Biomedical Optics Express 8(1): 57–67. OSA. doi:10.1364/BOE.8.000057.

Martins, Gabriela V., Ana P.M. Tavares, Elvira Fortunato and M. Goreti F. Sales. 2017. Paper-based sensing device for electrochemical detection of oxidative stress biomarker 8-hydroxy-2'-deoxyguanosine (8-OHdG) in point-of-care. Scientific Reports 7(1): 14558. doi:10.1038/s41598-017-14878-9.

Mi, Xiaona, Hui Li, Rong Tan, Bainian Feng and Yifeng Tu. 2021. The TDs/aptamer cTnI biosensors based on HCR and Au/Ti_3C_2-MXene amplification for screening serious patient in COVID-19 pandemic. Biosensors and Bioelectronics 192: 113482. doi:https://doi.org/10.1016/j.bios.2021.113482.

Palanisamy, S., S.K. Ramaraj, Shen-Ming Chen, Thomas C.K. Yang, Pan Yi-Fan, Tse-Wei Chen, Vijayalakshmi Velusamy, et al. 2017. A novel laccase biosensor based on laccase immobilized graphene-cellulose microfiber composite modified screen-printed carbon electrode for sensitive determination of catechol. Scientific Reports 7(1): 41214. doi:10.1038/srep41214.

Pan, Tung-Ming, Chih-Wei Wang, Wei-Che Weng, Chih-Chang Lai, Yu-Ying Lu, Chao-Yung Wang, I-Chang Hsieh, et al. 2022. Rapid and label-free detection of the troponin in human serum by a TiN-based extended-gate field-effect transistor biosensor. Biosensors and Bioelectronics 201: 113977. doi:https://doi.org/10.1016/j.bios.2022.113977.

Prabhakaran, Amrutha and Pranati Nayak. 2020. Surface engineering of laser-scribed graphene sensor enables non-enzymatic glucose detection in human body fluids. ACS Applied Nano Materials 3(1): 391–398. American Chemical Society. doi:10.1021/acsanm.9b02025.

Qi, Yingying, Fu-Rong Xiu, Minfeng Zheng and Baoxin Li. 2016. A simple and rapid chemiluminescence aptasensor for acetamiprid in contaminated samples: sensitivity, selectivity and mechanism. Biosensors and Bioelectronics 83: 243–249. doi:https://doi.org/10.1016/j.bios.2016.04.074.

Reynolds, Tess, Alexandre François, Nicolas Riesen, Michelle E. Turvey, Stephen J. Nicholls, Peter Hoffmann and Tanya M. Monro. 2016. Dynamic self-referencing approach to whispering gallery mode biosensing and its application to measurement within undiluted serum. Analytical Chemistry 88(7): 4036–4040. American Chemical Society. doi:10.1021/acs.analchem.6b00365.

Ribaut, Clotilde, Médéric Loyez, Jean-Charles Larrieu, Samia Chevineau, Pierre Lambert, Myriam Remmelink, Ruddy Wattiez, et al. 2017. Cancer biomarker sensing using packaged plasmonic optical fiber gratings: towards in vivo diagnosis. Biosensors and Bioelectronics 92: 449–456. doi:https://doi.org/10.1016/j.bios.2016.10.081.

Salazar, Pedro, Iñigo Fernández, Miriam C. Rodríguez, Alberto Hernández-Creus and José Luis González-Mora. 2019. One-step green synthesis of silver nanoparticle-modified reduced graphene oxide nanocomposite for H_2O_2 sensing applications. Journal of Electroanalytical Chemistry 855: 113638. doi:https://doi.org/10.1016/j.jelechem.2019.113638.

Sangubotla, Roopkumar and Jongsung Kim. 2021. Fiber-optic biosensor based on the laccase immobilization on silica-functionalized fluorescent carbon dots for the detection of dopamine and multi-color imaging applications in neuroblastoma cells. Materials Science and Engineering: C 122: 111916. doi:https://doi.org/10.1016/j.msec.2021.111916.

Shariati, Mohsen, Maryam Vaezjalali and Mahdi Sadeghi. 2021. Ultrasensitive and easily reproducible biosensor based on novel doped MoS_2 nanowires field-effect transistor in label-free approach for detection of hepatitis B virus in blood serum. Analytica Chimica Acta 1156: 338360. doi:https://doi.org/10.1016/j.aca.2021.338360.

Shariati, Mohsen, Mahdi Sadeghi and S.H. Reza Shojaei. 2022. Sensory analysis of hepatitis B virus DNA for medicinal clinical diagnostics based on molybdenum doped ZnO nanowires field effect transistor biosensor; a comparative study to PCR test results. Analytica Chimica Acta 1195: 339442. doi:https://doi.org/10.1016/j.aca.2022.339442.

Song, Pengfei, Hao Fu, Yongjie Wang, Cheng Chen, Pengfei Ou, Roksana Tonny Rashid, Sixuan Duan, et al. 2021. A microfluidic field-effect transistor biosensor with rolled-up indium nitride microtubes. Biosensors and Bioelectronics 190: 113264. doi:https://doi.org/10.1016/j.bios.2021.113264.

Soomro, Razium Ali, Sana Jawaid, Nazar Hussain Kalawar, Mawada Tunesi, Selcan Karakuş, Ayben Kilislioğlu and Magnus Willander. 2020. In-situ engineered MXene-TiO_2/ $BiVO_4$ hybrid as an efficient photoelectrochemical platform for sensitive detection of soluble CD44 proteins. Biosensors and Bioelectronics 166: 112439. doi:https://doi.org/10.1016/j.bios.2020.112439.

Sri, Smriti, Deepika Chauhan, G.B.V.S. Lakshmi, Alok Thakar and Pratima R. Solanki. 2022. MoS_2 nanoflower based electrochemical biosensor for TNF alpha detection in cancer patients. Electrochimica Acta 405: 139736. doi:https://doi.org/10.1016/j.electacta.2021.139736.

Stefano, Jéssica Santos, Luiz Ricardo Guterres e Silva, Raquel Gomes Rocha, Laís Canniatti Brazaca, Eduardo Mathias Richter, Rodrigo Alejandro Abarza Muñoz and Bruno Campos Janegitz. 2022. New conductive filament ready-to-use for 3D-printing electrochemical (Bio) sensors: towards the detection of SARS-CoV-2. Analytica Chimica Acta 1191: 339372. doi:https://doi.org/10.1016/j.aca.2021.339372.

Sukumar, Saranya, Agneeswaran Rudrasenan and Deepa Padmanabhan Nambiar. 2020. Green-synthesized rice-shaped copper oxide nanoparticles using caesalpinia bonducella seed extract and their applications. ACS Omega 5(2): 1040–1051. American Chemical Society. doi:10.1021/acsomega.9b02857.

Sun, Xiaolu, He Wang, Yannan Jian, Feifei Lan, Lina Zhang, Haiyun Liu, Shenguang Ge, et al. 2018. Ultrasensitive microfluidic paper-based electrochemical/visual biosensor based on spherical-like cerium dioxide catalyst for MiR-21 detection. Biosensors and Bioelectronics 105: 218–225. doi:https://doi.org/10. 1016/j.bios.2018.01.025.

Sun, Yujiao, Hui Jin, Xiaowen Jiang and Rijun Gui. 2020. Black phosphorus nanosheets adhering to thionine-doped 2D MOF as a smart aptasensor enabling accurate capture and ratiometric electrochemical detection of target MicroRNA. Sensors and Actuators B: Chemical 309: 127777. doi:https://doi. org/10.1016/j.snb.2020.127777.

Sun, Yuezhen, Xiaoxia Guo, Yarien Moreno, Qizhen Sun, Zhijun Yan and Lin Zhang. 2022. Sensitivity adjustable biosensor based on graphene oxide coated excessively tilted fiber grating. Sensors and Actuators B: Chemical 351: 130832. doi:https://doi.org/10.1016/j.snb.2021.130832.

Thanihaichelvan, M., S.N. Surendran, T. Kumanan, U. Sutharsini, P. Ravirajan, R. Valluvan and T. Tharsika. 2022. Selective and electronic detection of COVID-19 (Coronavirus) using carbon nanotube field effect transistor-based biosensor: a proof-of-concept study. Materials Today: Proceedings 49: 2546–49. doi:https://doi.org/10.1016/j.matpr.2021.05.011.

Ullah, Najeeb, Wei Chen, Beenish Noureen, Yulan Tian, Liping Du, Chunsheng Wu and Jie Ma. 2021. An electrochemical Ti_3C_2Tx aptasensor for sensitive and label-free detection of marine biological toxins. Sensors. doi:10.3390/s21144938.

Wang, Zhongrong and Yunfang Jia. 2018. Graphene solution-gated field effect transistor DNA sensor fabricated by liquid exfoliation and double glutaraldehyde cross-linking. Carbon 130: 758–767. doi:https://doi.org/10. 1016/j.carbon.2018.01.078.

Wang, Rui, Jiu-Ju Feng, Yadong Xue, Liang Wu and Ai-Jun Wang. 2018a. A label-free electrochemical immunosensor based on AgPt nanorings supported on reduced graphene oxide for ultrasensitive analysis of tumor marker. Sensors and Actuators B: Chemical 254: 1174–1181. doi:https://doi.org/10.1016/j.snb. 2017.08.009.

Wang, Yingyi, Shanshan Wang, Chunsong Lu and Xiaoming Yang. 2018b. Three kinds of DNA-directed nanoclusters cooperating with graphene oxide for assaying mucin 1, carcinoembryonic antigen and cancer antigen 125. Sensors and Actuators B: Chemical 262: 9–16. doi:https://doi.org/10.1016/j. snb.2018.01.235.

Wang, Xu-dong and Otto S. Wolfbeis. 2020. Fiber-optic chemical sensors and biosensors (2015–2019). Analytical Chemistry 92(1): 397–430. American Chemical Society. doi:10.1021/acs.analchem.9b04708.

Wang, Song, Lantian Su, Lumei Wang, Dongwei Zhang, Guoqing Shen and Yun Ma. 2020. Colorimetric determination of carbendazim based on the specific recognition of aptamer and the poly-diallyldimethylammonium chloride aggregation of gold nanoparticles. Spectrochimica Acta Part A: Molecular and Biomolecular Spectroscopy 228: 117809. doi:https://doi.org/10.1016/j.saa.2019.117809.

Wang, Zhaoliang, Haiyang Yu and Zheng Zhao. 2021a. Silk fibroin hydrogel encapsulated graphene filed-effect transistors as enzyme-based biosensors. Microchemical Journal 169: 106585. doi:https://doi.org/10.1016/j.microc.2021.106585.

Wang, Peilin, Yixin Nie, Yu Tian, Zihui Liang, Shuping Xu and Qiang Ma. 2021b. A whispering gallery mode-based surface enhanced electrochemiluminescence biosensor using biomimetic antireflective nanostructure. Chemical Engineering Journal 426: 130732. doi:https://doi.org/10.1016/j.cej.2021.130732.

Wang, Bingfang, Yuanyuan Luo, Lei Gao, Bo Liu and Guotao Duan. 2021c. High-performance field-effect transistor glucose biosensors based on bimetallic Ni/Cu metal-organic frameworks. Biosensors and Bioelectronics 171: 112736. doi:https://doi.org/10.1016/j.bios.2020.112736.

Wang, Junxia, Mengqi Lv, Hehuan Xia, Jialei Du, Yiwei Zhao, Hao Li and Zhongyin Zhang. 2021d. Minimalist design for a hand-held SARS-Cov-2 sensor: peptide-induced covalent assembly of hydrogel enabling facile fiber-optic detection of a virus marker protein. ACS Sensors 6(6): 2465–2471. American Chemical Society. doi:10.1021/acssensors.1c00869.

Wei, Xiaofeng, Jialei Guo, Huiting Lian, Xiangying Sun and Bin Liu. 2021. Cobalt metal-organic framework modified carbon cloth/paper hybrid electrochemical button-sensor for nonenzymatic glucose diagnostics. Sensors and Actuators B: Chemical 329: 129205. doi:https://doi.org/10.1016/j.snb.2020.129205.

Wu, Jixuan, Bo Wang, Binbin Song, Mingqiang Qiao, Bo Liu, Hao Zhang, Wei Lin, et al. 2021. Bioimmunoassay based on hydrophobin HGFI self-assembled whispering gallery mode optofluidic microresonator. Sensors and Actuators A: Physical 319: 112545. doi:https://doi.org/10.1016/j.sna.2021.112545.

Xia, Jianfei, Xiyue Cao, Zonghua Wang, Min Yang, Feifei Zhang, Bing Lu, Feng Li et al. 2016. Molecularly imprinted electrochemical biosensor based on chitosan/ionic liquid–graphene composites modified electrode for determination of bovine serum albumin. Sensors and Actuators B: Chemical 225: 305–311. doi:https://doi.org/10.1016/j.snb.2015.11.060.

Xia, Tianzi, Guangyan Liu, Junjie Wang, Shili Hou and Shifeng Hou. 2021. MXene-based enzymatic sensor for highly sensitive and selective detection of cholesterol. Biosensors and Bioelectronics 183: 113243. doi:https://doi.org/10.1016/j.bios.2021.113243.

Xiao, Peng, Zhen Sun, Yan Huang, Wenfu Lin, Yuchen Ge, Ruitao Xiao, Kaqiang Li, et al. 2020. Development of an optical microfiber immunosensor for prostate specific antigen analysis using a high-order-diffraction long period grating. Optics Express 28(11): 15783–15793. OSA. doi:10.1364/OE.391889.

Xiong, Can, Tengfei Zhang, Weiyu Kong, Zhixiang Zhang, Hao Qu, Wei Chen, Yanbo Wang, et al. 2018. ZIF-67 derived porous Co_3O_4 hollow nanopolyhedron functionalized solution-gated graphene transistors for simultaneous detection of glucose and uric acid in tears. Biosensors and Bioelectronics 101: 21–28. doi:https://doi.org/10.1016/j.bios.2017.10.004.

Xu, Xiaolin, Min Wei, Yuanjian Liu, Xu Liu, Wei Wei, Yuanjian Zhang and Songqin Liu. 2017. A simple, fast, label-free colorimetric method for detection of telomerase activity in urine by using hemin-graphene conjugates. Biosensors and Bioelectronics 87: 600–606. doi:https://doi.org/10.1016/j.bios.2016.09.005.

Xu, Jiayao, Ming Shi, Huakui Huang, Kun Hu, Wenting Chen, Yong Huang and Shulin Zhao. 2018. A fluorescent aptasensor based on single oligonucleotide-mediated isothermal quadratic amplification and graphene oxide fluorescence quenching for ultrasensitive protein detection. Analyst 143(16): 3918–3925. The Royal Society of Chemistry. doi:10.1039/C8AN01032C.

Xu, Shicai, Tiejun Wang, Guofeng Liu, Zanxia Cao, Ludmila A. Frank, Shouzhen Jiang, Chao Zhang, et al. 2021a. Analysis of interactions between proteins and small-molecule drugs by a biosensor based on a graphene field-effect transistor. Sensors and Actuators B: Chemical 326: 128991. doi:https://doi.org/10.1016/j.snb.2020.128991.

Xu, Yichao, Meng Xiong and Hui Yan. 2021b. A portable optical fiber biosensor for the detection of zearalenone based on the localized surface plasmon resonance. Sensors and Actuators B: Chemical 336: 129752. doi:https://doi.org/10.1016/j.snb.2021.129752.

Yang, Xiao, Minghui Feng, Jianfei Xia, Feifei Zhang and Zonghua Wang. 2020. An electrochemical biosensor based on AuNPs/Ti_3C_2 MXene three-dimensional nanocomposite for MicroRNA-155 detection by exonuclease III-aided cascade target recycling. Journal of Electroanalytical Chemistry 878: 114669. doi:https://doi.org/10.1016/j.jelechem.2020.114669.

Ye, Tai, Wenxiang Yin, Nianxin Zhu, Min Yuan, Hui Cao, Jingsong Yu, Zongqin Gou, et al. 2018. Colorimetric detection of pyrethroid metabolite by using surface molecularly imprinted polymer. Sensors and Actuators B: Chemical 254: 417–423. doi:https://doi.org/10.1016/j.snb.2017.07.132.

Zamzami, Mazin A., Gulam Rabbani, Abrar Ahmad, Ahmad A. Basalah, Wesam H. Al-Sabban, Saeyoung Nate Ahn and Hani Choudhry. 2022. Carbon nanotube field-effect transistor (CNT-FET)-based biosensor for rapid detection of SARS-CoV-2 (COVID-19) surface spike protein S1. Bioelectrochemistry 143: 107982. doi:https://doi.org/10.1016/j.bioelechem.2021.107982.

Zhang, Chan, Pengfei Du, Zejun Jiang, Maojun Jin, Ge Chen, Xiaolin Cao, Xueyan Cui, et al. 2018. A simple and sensitive competitive bio-barcode immunoassay for triazophos based on multi-modified gold nanoparticles and fluorescent signal amplification. Analytica Chimica Acta 999: 123–131. doi:https://doi.org/10.1016/j.aca.2017.10.032.

Zhang, Xinmeng, Zixuan Mao, Yuanxiao Zhao, Yuanting Wu, Changqing Liu and Xiufeng Wang. 2020. Highly sensitive electrochemical sensing platform: carbon cloth enhanced performance of Co_3O_4/rGO nanocomposite for detection of H_2O_2. Journal of Materials Science 55(13): 5445–5457. doi:10.1007/s10853-020-04393-0.

Zhao, Shuai, Tuoyu Zhou, Aman Khan, Zhengjun Chen, Pu Liu and Xiangkai Li. 2022. A novel electrochemical biosensor for bisphenol a detection based on engineered escherichia coli cells with a surface-display of tyrosinase. Sensors and Actuators B: Chemical 353: 131063. doi:https://doi.org/10.1016/j.snb.2021.131063.

Zhou, Zhuoyue, Zhao Yang, Li Xia and Houjin Zhang. 2022. Construction of an enzyme-based all-fiber SPR biosensor for detection of enantiomers. Biosensors and Bioelectronics 198: 113836. doi:https://doi.org/10.1016/j.bios.2021.113836.

Index

For Product Safety Concerns and Information please contact our EU
representative GPSR@taylorandfrancis.com
Taylor & Francis Verlag GmbH, Kaufingerstraße 24, 80331 München, Germany

www.ingramcontent.com/pod-product-compliance
Lightning Source LLC
Chambersburg PA
CBHW060402220326
41598CB00023B/2993

9 781032 573533